Springer Theses

Recognizing Outstanding Ph.D. Research

For further volumes:
http://www.springer.com/series/8790

Aims and Scope

The series "Springer Theses" brings together a selection of the very best Ph.D. theses from around the world and across the physical sciences. Nominated and endorsed by two recognized specialists, each published volume has been selected for its scientific excellence and the high impact of its contents for the pertinent field of research. For greater accessibility to non-specialists, the published versions include an extended introduction, as well as a foreword by the student's supervisor explaining the special relevance of the work for the field. As a whole, the series will provide a valuable resource both for newcomers to the research fields described, and for other scientists seeking detailed background information on special questions. Finally, it provides an accredited documentation of the valuable contributions made by today's younger generation of scientists.

Theses are accepted into the series by invited nomination only and must fulfill all of the following criteria

- They must be written in good English.
- The topic should fall within the confines of Chemistry, Physics, Earth Sciences, Engineering and related interdisciplinary fields such as Materials, Nanoscience, Chemical Engineering, Complex Systems and Biophysics.
- The work reported in the thesis must represent a significant scientific advance.
- If the thesis includes previously published material, permission to reproduce this must be gained from the respective copyright holder.
- They must have been examined and passed during the 12 months prior to nomination.
- Each thesis should include a foreword by the supervisor outlining the significance of its content.
- The theses should have a clearly defined structure including an introduction accessible to scientists not expert in that particular field.

Philip G. Born

Crystallization of Nanoscaled Colloids

Doctoral Thesis accepted by
the Saarland University, Germany

Research carried out at the Leibniz Institute
for New Materials, Saarbrücken, Germany

 Springer

Author
Dr. Philip G. Born
INM – Leibniz Institute for New Materials
Saarbrücken
Germany

Present Address
German Aerospace Center
DLR Institute of Materials Physics in Space
  (DLR-MP)
Cologne
Germany

Supervisor
Dr. Tobias Kraus
INM – Leibniz Institute for New Materials
Saarbrücken
Germany

ISSN 2190-5053 ISSN 2190-5061 (electronic)
ISBN 978-3-319-03296-2 ISBN 978-3-319-00230-9 (eBook)
DOI 10.1007/978-3-319-00230-9
Springer Cham Heidelberg New York Dordrecht London

Printed on acid-free paper

Springer is part of Springer Science+Business Media (www.springer.com)

Parts of this thesis have been published in the following journal articles:

P. Born, E. Murray und T. Kraus. *Temperature-induced particle self-assembly.* J. Phys. Chem. Solids *71*, 95–99 (2010).

E. Murray, P. Born, A. Weber and T. Kraus. *Synthesis of Monodisperse Silica Nanoparticles Dispersable in Non-Polar Solvents.* Adv. Eng. Mat. *12*, 374–378 (2010).

E. Murray, P. Born and T. Kraus. *Colloidal Polymer Patterning* in: Generating Micro- and Nanopatterns on Polymeric Materials, Wiley–VCH, 2011.

M. Schulzendorf, C. Cavelius, P. Born, E. Murray and T. Kraus. *Biphasic Synthesis of Au@SiO2 Core–Shell Particles with Stepwise Ligand Exchange.* Langmuir, *27*, 727–732 (2011).

P. Born, S. Blum, A. Munoz and T. Kraus. *Role of the Meniscus Shape in Large-Area Convective Particle Assembly.* Langmuir, *27*, 8621–8633 (2011).

P. Born, A. Munoz, C. Cavelius and T. Kraus. *Crystallization Mechanisms in Convective Particle Assembly.* Langmuir, *28*, 8300–8308 (2012).

T. Geyer*, P. Born*, and T. Kraus (*equal contribution). *Switching between Crystallization and Amorphous Agglomeration of Alkyl Thiol-Coated Gold Nanoparticles.* Phys. Rev. Let. *109*, 128302 (2012).

P. Born, and T. Kraus. *Ligand-dominated temperature dependence of agglomeration kinetics and morphology in alkyl thiol-coated gold nanoparticles.* submitted (2013).

Supervisor's Foreword

In this thesis, Philip G. Born investigates colloidal sub-micron particles. Suspensions of such particles are convenient agents to introduce small length scales into materials, for example, by mixing them with a polymer. Confinement effects and large interfaces endow the resulting composite material with useful properties: color, mechanical toughness, and many others, depending on the particle type. Today, particle-containing materials are used in optical materials, for electronics and photovoltaics, in protective coatings and in biomedicine, to name only a few.

Particles are not only useful additives for materials. Dispersions of monodispersed particles resemble atoms and molecules. The particles are in constant motion, "react" to dimers, and exhibit phase transitions. In contrast to atoms, however, their interactions can be tuned. Individual particles are also easier to observe than atoms. Charge-stabilized particles have been used as models to study nucleation and crystallization phenomena for decades. Modern nanoparticles add a wide variety of interactions, with interaction lengths that can span multiple particle diameters.

This thesis is concerned with superstructures that particles form when they precipitate. Philip G. Born describes experiments to reconstruct the particles' motion during this process. The question he asks is of both technological and fundamental interest: what governs the agglomerates' geometries? If the particles are almost identical, what keeps them from crystallizing? Such questions are the basis for smooth, uniform, and flat particle films. If particle films are to compete with vacuum-deposited thin films in applications, we must answer them.

Several in situ methods were used in this thesis to observe the assembling particles. Optical microscopy tracked marked particles moving in complex flows, dynamic light scattering indicated the mobility of small metal particles. Both methods are well-established, but in this thesis, the particles' interactions were tuned during observation to establish a link between interactions and agglomerate morphologies.

To change the interactions of larger particles, Philip tuned their solvent's evaporation rate. Evaporation caused convection that hydrodynamically coupled the particles. A minimum temperature is required to create sufficient convection and deposit crystalline particle layers. Just as particles only sediment and crystallize in barometric equilibrium if the gravitational field overcomes diffusion, the hydrodynamic forces in non-equilibrium deposition must outweigh diffusion. Unlike gravitation, convection is temperature-dependent, which resulted in a critical temperature.

In smaller particles, Philip exploited the ligand shell to tune interactions. At low temperatures, the ligands formed solid shells around the particles. At higher temperatures, the shells melted. The transition in the ligand shell translated into changing interactions and agglomerate morphologies. A minimum temperature for crystallization emerges again, but its origin is entirely different from that above. Here, subtle changes in the short-range interactions between nanoparticles prevent or allow crystallization. This effect has not been reported before.

The dispersions discussed in this thesis are complex systems: Particles are more complex than atoms, suspensions with solvents and interfaces add further complexity. Is it reasonable to use particles as models for atoms—and can we use concepts that were developed for atoms to better understand the behavior of particles? Philip Born's thesis shows that the analogy is not the final answer, but a useful starting point.

Saarbrücken, Feburary 2013   Dr. Tobias Kraus

Acknowledgments

Many people contributed to the completion of this thesis. If not all of them are recognized correspondingly, I want to apologize. I acknowledge them all!

First of all I thank my Ph.D. supervisor, Prof. Dr. Eduard Arzt. His efforts in renewing and broadening the research at the INM led to the foundation of junior research groups and the treatment of the research presented here.

I thank Dr. Tobias Kraus for taking the risk of starting a junior research group in a changing institute. The fascinating topic of self-assembly of colloidal particles would not have captured me without him! All other former and present group members have to be acknowledged for companionship and support.

Prof. Dr. Wenz is kindly acknowledged for the scientific supervision of this thesis. His moral support significantly helped in completing this thesis.

Unpayable were the coffee breaks with Christoph Erlenkämper, especially scientifically. I enjoyed his fun with "the physics behind it" very much, thanks a lot!

The colleagues at the INM have to be thanked in great generality. I had a great time, and the scientific spirit at the institute is exemplary. In extracts, I thank the team of the physical analytics group for the introduction to electron microscopy, the workshop for magically constructing working setups from my sketchy plans, the colleagues suffering with me through the nights at the synchrotron beamlines, Petra Herbeck-Engel for DSC- and Raman- and Claudia Fink-Straube for ICP-AES-measurements, the folks from the nanotribology group whom I motivated to incredible successful friction measurements, and of course the students from the coffee- and kicker-gatherings and the Ph.D. seminar for many breaks.

Completion of this thesis would not have been possible without the long-lasting support of my parents. I did not always acknowledge this properly, but at this point I can do it at least rudimentary.

The most affectionate gratefulness applies to my wife, who suffered most due to this thesis. Thank you for the patience and endurance! I am looking forward to many nice days!

Contents

Abbreviations and Symbols

$1/R_\parallel$	Curvature of the meniscus parallel to the substrate
$1/R_\perp$	Curvature of the meniscus perpendicular to the substrate
χ	Flory–Huggins-parameter
η	Solvent viscosity
γ	Interfacial energy
γ_{lg}	Liquid–gas interfacial energy
γ_{sg}	Solid–gas interfacial energy
γ_{sl}	Solid–liquid interfacial energy
κ	Debye screening length
κ^{-1}	Capillary length
λ	Wavelength
ϕ	Particle volume fraction in suspension
ρ_l	Liquid density
σ	Root mean square deviation
Θ	Contact angle
Θ_d	Dynamic contact angle
Θ_s	Static contact angle
A	Hamaker coefficient
D	Diffusion coefficient
d	Particle surface separation
DLS	Dynamic light scattering
DSC	Differential scanning calorimetry
DWS	Diffusing wave spectroscopy
$eDLS$	Echoed dynamic light scattering
F	Helmholtz free energy
F_i	Capillary forces among particles protruding through a liquid surface
$FT\text{–}IR$	Fourier transform-infrared spectroscopy
$FWHM$	Full width half maximum
g	Gravitational acceleration
$g_{(s)}$	Radial distribution function
H	Distance between substrate and blade
h_f	Height of the particle film
$I(q)$	Scattered Intensity

$ICP\text{–}AES$	Inductively coupled plasma–atomic emission spectroscopy
IR	Infrared spectroscopy
J_e	Evaporative flow of the solvent
j_e	Evaporative flux density of solvent
J_p	Particle flux to the film growth front
j_p	Particle flux density
J_w	Solvent flow from the bulk suspension
j_w	Solvent flux density
k	Spring constant
k_{11}	Rate constant of coagulation of primary particles in suspension
k_{11}^{Smol}	Rate constant of fast or Smoluchowskian agglomeration
L	Polymer chain length
l	Linker segment length of polymer chain
l_{buffer}	Buffer length
l_{dep}	Depinning length of a meniscus
l_e	Length of wet particle film, 'evaporation length'
n	Number of primary particles in solution
n_p	Polymer chain number density
p	Pressure
q	Scattering vector
r	Particle radius
$R_{\parallel}$	Radius of curvature parallel to substrate
$R_{\perp}$	Radius of curvature perpendicular to substrate
S	Entropy
S	Center-to-center particle distance
$S(q)$	Static structure factor
$SAXS$	Small-angle X-ray scattering
T	Absolute temperature
t	Time
T_A	Temperature of agglomeration onset
t_{buffer}	Buffer time
T_c	Temperature above which suspended particles agglomerate into colloidal crystals
T_m	Melting temperature
TEM	Transmission electron microscopy
U	Internal energy
$V(s)$	Interparticle potential
V_{charge}	Interaction potential generated by particle surface charges
v_{cl}	Velocity of contact line with respect to substrate
v_f	Particle film growth rate
V_{mas}	Height of potential barrier in charge-stabilized suspensions
V_{min}	Depth of the effective interparticle potential
v_m	Maximal possible contact line velocity
V_{steric}	Steric interaction potential
V_{vdW}	Van der Waals interaction potential, dispersion interaction

W	Stability ratio
w_f	Width of the deposited particle film
x	Coordinate axis parallel to substrate from film growth front into bulk suspension
y	Coordinate axis parallel to substrate and parallel to film growth front
z	Coordinate axis perpendicular to substrate and perpendicular to film growth front

Chapter 1
Introduction

Crystals made out of colloidal nanoparticles (Fig. 1.1a, b) introduce novel functionalities to condensed matter: their properties stem both from properties of individual nanoparticles and from collective properties. Examples for nanoparticle properties contributing to the functionality of colloidal crystals are plasmon frequencies of metal nanoparticles, band gap tuning in semiconducting nanoparticles and super-paramagnetic behavior of magnetic nanoparticles [1, 2]. Examples for the collective properties are photonic bandgaps due to the three-dimensional periodicity on the order of the wavelength of light [3, 4] and the generation of minibands as collective states of ordered arrays of quantum dots [5], but also selectively electric or thermal conductivity due to the high connectivity in close-packed structures. The recent availability of nanoparticles with narrow shape, size and composition dispersity nourishes the fast introduction of colloidal crystals into applications such as optical circuits, photovoltaics and thermoelectrics.

The production of functional materials using nanoparticles as building blocks requires a paradigm shift. In a conventional approach each particle would be positioned like a brick in a wall. Instruments like optic tweezers or scanning probe microscopes can provide this precision. However, even if it would take only a tenth of a second to put a single particle in its place, it would take more than 900 years to fill a square centimeter with a monolayer of hexagonally packed nanoparticles with a radius of 10 nm. Practical production of colloidal crystals thus can only proceed through self-assembly of the particles. This initiated a search for suitable approaches and conditions to induce self-assembly of particles in the past decade. In a long-term perspective even complex structures like conducting paths and transistors can be expected to be fabricated from particles by self-assembly. Self-assembly could then be used to support established technology in the generation of nanostructures.

However, the application of crystals made out of colloidal nanoparticles is limited today by several factors. Reproducible preparation of large crystals beyond tens of micrometers with controlled dimensions and small packing defect densies remain challenging. Controlling the morphology of colloidal crystals and their placement, important factors for device integration, also are not yet solved tasks. The problems

P. G. Born, *Crystallization of Nanoscaled Colloids*, Springer Theses,
DOI: 10.1007/978-3-319-00230-9_1, © Springer International Publishing Switzerland 2013

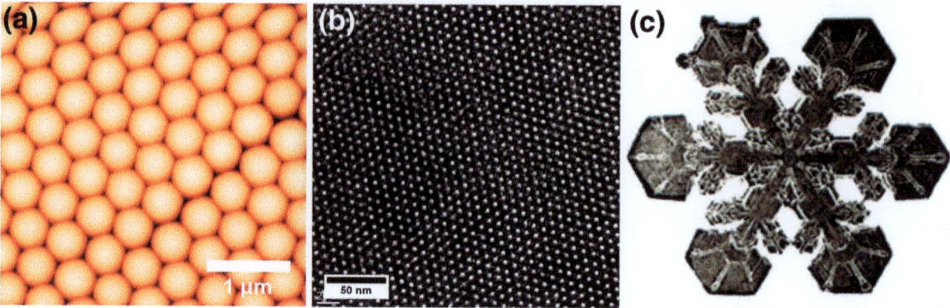

Fig. 1.1 a and **b** colloidal crystals. **a** Atomic force micrograph of convectively assembled 500 nm polystyrene particles. **b** Transmission electron micrograph of precipitated crystal of 6 nm gold particles. **c** A snowflake [6], exemplifying the high complexity of crystal morphology obtainable beyond the dense packing of the constituting molecules

with optimization of crystal formation arise from the limited access to in-situ information on the crystallization process and, to some extent, from uncertainties about the dominant interaction among the particles.

Consequently, the fundamental questions to be solved in research on crystallization of nanoscaled colloids are: how can large defect-free colloidal crystals with defined dimensions be grown, and how can structure formation be controlled beyond dense packing of hard spheres? Growing colloidal crystals by rational choice of external conditions into defined morphologies might render colloidal self-assembly a competitor to techniques such as lithography or molding. The example of a snowflake (Fig. 1.1c) suggests that complex structures are not only defined by the minimal free energy of the constituting objects but also by the kinetics of the assembly process, which can be set by external conditions like temperature and humidity.

The selection or development of routes towards the crystallization of nanoscaled colloids require a basic understanding of interactions present in colloids, comparison of present approaches to this problem and clarity of the used terms. Terms such as self-organization, self-assembly, colloids and crystals are used in distinct fields including biology, chemistry, physics and material science, but also in psychology, ecology and computer science, with varying meaning and focus. To avoid misunderstandings, the main terms are defined in the following section.

1.1 Definitions

Colloids are defined as "small objects suspended in liquids" [7–10]. The small objects can be solid particles, macromolecules or molecular assemblies. Examples for the first category are gold nanoparticles or clays, polymer latices or polymer chains fall in the second category, and micelles are representative for the last category. "Small" implies characteristic sizes of the objects from few nanometers up to several micrometers. The lower limit requires that the objects are significantly larger than the solvent molecules. The upper limit ensures that the objects perform BROWNIAN motion and gravitational settling is not dominant.

The lack of sharp boundaries creates gray zones. The upper limit is used to distinguish colloids from granular matter. Granular objects do not perform BROWNIAN motion, and external forces like gravitation and contact forces like friction are dominant [11]. But both colloid and granulate are only idealized extremes of the same class of many-body systems interacting by short-range interactions but not by interactions arising from overlapping electron wave-functions. This differentiates colloids and granulates from atoms, which have short-ranged, often highly directional and specific strong interactions, and from systems with long-ranged interactions like planets or plasmas that interact through gravity or unscreened electrostatic interactions. The underlying similarities between colloids and granulates and the continuous transition between them suggest that phenomena dominant in one system may still play a subtle role in the other.

The restriction of the experiments presented here to solid particles suspended in liquids render the terms particle suspension, colloidal suspension and colloid synonymous throughout this work.

Nanoscaled colloids are colloids with the diameters of the suspended particles below 100 nm.

Crystalline packing is defined as an "infinite [periodic] repetition of identical subunits in three dimensions" in atomic systems [12]. This definition cannot be used unmodified for colloids for three reasons. First, colloids often consist of a very limited number of constituting objects, thus they cannot form infinite repetitions. Second, the constituting objects of colloids exhibit finite dispersity in size, shape and sometimes even composition, thus identical subunits cannot be formed by colloids. Finally, two-dimensional regular particle packings may have common genesis and properties with three-dimensional packings, and should be included in the definition.

A robust definition of colloidal crystals is still lacking; therefore a phenomenological definition is used in the presented work. A structure or packing made out of particles is called a crystal when it exhibits an obvious six-fold symmetry and a periodicity over several particle diameters in imaging techniques such as electron microscopy or BRAGG-peaks in scattering techniques such as small-angle X-ray scattering. The restriction throughout the experiments described in this thesis to spherical particles with low size and shape dispersity render this definition useful. In future work a more general definition of colloidal crystals should be developed.

Crystalline packings, crystals and regular packings are used as synonyms throughout the text.

Amorphous packings are not crystalline packings. They do not exhibit sixfold symmetry and periodicity over several particle diameters or BRAGG-peaks. Amorphous packings and irregular packings are used as synonyms.

Structure formation denotes all processes that lead to deviations from uniform distributions. The formation of a colloidal crystal is structure formation, as the colloidal particles are arranged in a crystalline lattice instead of being distributed throughout the suspension.

Self-organization processes are structure forming processes that rely on the interactions of the constituting objects. In case a colloidal crystal is not formed by putting each particle in its lattice site extrinsically but by the interactions among the objects, it is a self-organized structure. Self-organization is meant to cover all structure forming processes relying on the interactions of the constituting objects including self-assembly, whether they form dynamic or static structures and whether may the structure formation takes place during equilibration or in processes far from equilibrium.

Self-assembly focuses on the assembly of materials by self-organization processes. Following the idea of 'self-assembly' or 'static self-assembly' given by WHITESIDES and co-workers [13, 14], which represents a chemist's perspective on the topic, 'assembled' structures like liquid crystals, phase separated metal alloys and colloidal crystals can be distinguished from other 'organized' structures like a school of fishes, chemical oscillations during the Belousov–Zhabotinsky reaction or the morphogenesis of multicellular organisms. Self-assembly is therefore a subset of self-organization.

The following section gives a brief overview over mechanisms acting in colloids, from which the basic principles of colloidal crystallization are deduced. Consequently, this allows to categorize results on colloidal crystallization in the literature and, in the last section, rationalizing the experiments in this work in the last section.

1.2 Interactions and Mechanisms in Colloids

Comprehensive textbooks on basic principles in colloids are available, for example by W. B. RUSSEL, D. A. SAVILLE and W. R. SCHOWALTER [7], J. LYKLEMA [8], P. N. PUSEY [9], D. F. EVANS and H. WENNERSTRÖM [10] and J. N. ISRAELACHVILI [15]. To aid the reading and rationalize the experiments, a brief overview of colloidal interactions and mechanisms is given at this point.

The most prominent feature of colloids is the constant erratic BROWNIAN motion of the components. It is caused by fluctuating momentum transfer from solvent molecules to the suspended particles. The root-mean-square displacement motion $\langle r^2 \rangle^{\frac{1}{2}}$ of the particles in time t due to the random walk can be quantified using the STOKES- EINSTEIN diffusion coefficient $D_0 = kT/6\pi\eta r$ [7]:

$$\langle r^2 \rangle^{\frac{1}{2}} = \left(\frac{2kT}{6\pi\eta r} \cdot t \right)^{\frac{1}{2}} = (2\,D_0 \cdot t)^{\frac{1}{2}}, \qquad (1.1)$$

where k is BOLTZMANN's constant, T the absolute temperature, η the solvent viscosity and r the particle radius. BROWNIAN motion is a crucial factor in particle assembly. The mass transport from a source, the bulk colloid, to a drain, the pattern, can be accomplished by diffusion. After assembly, the particles have to be trapped to prevent further random motion.

The exchange of energy and momentum between the liquid and the particles occurs on a timescale short compared to that of significant motions of the particles. The solvent thus can be approximated to simply provide a heat bath that defines the temperature. The BOLTZMANN-distribution $P(v)$ applies to the particle velocities v [8]:

$$P(v) = \sqrt{\frac{2}{\pi}} \left(\frac{m}{kT}\right)^{\frac{3}{2}} v^2 e^{-\frac{mv^2}{2kT}}, \tag{1.2}$$

where m denotes the mass of the particles. The standard deviation of the BOLTZMANN distribution is proportional to $1/m$. Already very fine nanoparticles of $1\,\mathrm{nm}^3$ volume contain approximately 100 atoms, and the particle energy distribution width becomes a 100th of the width of the corresponding atomic distribution. In practical colloids, virtually all particles have the same velocity and energy. Effects like sublimation and evaporation thus will not play significant roles in colloidal particle assemblies. The probability of a particle having enough energy to overcome the attractive interactions in an assembly is very low unless the whole assembly is heated to the melting or boiling temperature.

Besides providing energy and mobility, the solvent dampens the motion of the particles. This is expressed in the STOKES drag, the force experienced by a sphere of radius r moving with velocity v through a solvent at rest [7]:

$$F(v) = 6\pi\eta\,r\,v. \tag{1.3}$$

In addition, the particles interact with each other. We restrict the particle interactions to attractive and repulsive potentials. Velocity-dependent interactions like hydrodynamic interactions are neglected, as they are expected not to effect the structure of static particle packings.

Attraction among the particles is due to the VAN DER WAALS- or dispersion interactions among the particles. The magnitude of the attraction depends on the difference in dielectric permittivities of the interacting bodies and the surrounding media, which is approximated by the HAMAKER coefficient A. For equal spheres, the VAN DER WAALS-attraction takes the form [16]:

$$V_{vdW} = -\frac{1}{3} A \left(\frac{r^2}{s^2 - 4r^2} + \frac{r^2}{s^2} + \frac{1}{2} \ln\left(1 - 4\frac{r^2}{s^2}\right)\right), \tag{1.4}$$

where s is the center-to-center distance between particles of radius r. Dispersion forces can be repulsive, but for chemically equal particles as used in this work they generate an omnipresent attraction. Two important approximations of the VAN DER WAALS-attraction can be given. For particle surface separation $d = s - 2r$ small compared to the particle radius r, the potential takes the form

$$V_{vdW} = -\frac{A}{6}\frac{r}{d}. \tag{1.5}$$

For particles at large separations, $s \gg r$, the potential takes the well-known s^{-6} dependency:

$$V_{vdW} = -\frac{16A}{9}\frac{r^6}{s^6}.$$ (1.6)

A particle suspension has to be stabilized against this omnipresent attraction to prevent agglomeration and precipitation of the particles. Two basic principles have been developed to prevent agglomeration. First, charges on the particle surfaces create a repulsive interaction. In the approximation of saturated surfaces the electrostatic potential of a sphere screened by counterions in the solvent can be approximated as [7]

$$V_{charge} \approx \Psi_0 \frac{r}{s} e^{-\kappa \cdot d}.$$ (1.7)

Ψ_0 is the potential at the particle surface ($d = 0$) and κ^{-1} indicates the thickness of the diffuse layer of ions around the charged particles. $\kappa = 2e^2 z^2 n_i / \epsilon\epsilon_0 kT$ is called the DEBYE screening length, with e denoting the elementary charge, ϵ and ϵ_0 the relative and the vacuum permittivity and z and n_i the valence and concentration of the ionic species, respectively. Importantly, the screened electrostatic repulsions do not diverge for $s \to r$ or alternatively $d \to 0$ as the VAN DER WAALS-attraction does (Eq. 1.5). The electrostatic stabilization thus only creates a kinetic barrier against agglomeration, and the particles overcome the repulsive barrier with a probability $P_a \approx e^{-\frac{V_{charge,max}}{kT}}$ upon collision, where $V_{charge,max}$ is the height of the repulsive barrier.

Second, polymeric chains grafted on the particle surfaces can create repulsion between the particles. The repulsion arises from steric effects of the chains. A close proximity of two particles would increase the chain density between the particles. This restricts the motion of the chains and lowers the entropy, thus raises the free energy of the particles. The repulsive potential generated between two polymer coated particles per unit area can be written in two approximative forms as [17]

$$V_{steric} = \frac{N_p kT}{A_s}\left[\frac{\pi^2 l L}{6d^2} + \ln d\left(\frac{3}{8\pi l L}\right)^{\frac{1}{2}}\right] \qquad \frac{d^2}{lL} \leq 3 \qquad (1.8)$$

$$= \frac{N_p kT}{A_s} 2\, e^{\frac{-3d^2}{2lL}} \qquad\qquad\qquad \frac{d^2}{lL} \geq 3, \qquad (1.9)$$

where N_p/A_s denotes the number of polymer chains per surface area, and l and L the polymer's segment length and chain length, respectively. Importantly, the steric repulsion diverges faster than the VAN DER WAALS-attraction for short distances (Eq. 1.5). At short distances sterically stabilized particles therefore are repulsive. At long distances the steric repulsion decays exponentially, with little interaction above the chain length of the polymer, and the dispersion interactions may become prevailing, depending on the exact parameter values.

In the derivation of Eq. 1.9 random-flight polymers are assumed. This holds true only for certain solvents and temperatures. The influence of the solvent can be approximated with the FLORY- HUGGINS-parameter χ [18],

$$\chi = \frac{z}{kT}\left(u_{sp} - \frac{u_{ss} + u_{pp}}{2}\right).$$ (1.10)

The coordination number z counts the solvent molecules or segments each polymer segment interacts with. The parameter u_{ss}, u_{pp} and u_{sp} indicate the interactions among solvent molecules (ss), among segments in the polymer chain (pp) and between solvent molecules and segments (sp) in units of kT. A value of χ of 1/2 indicates random-flight polymers, the respective temperature T and solvent are termed θ-temperature and θ-solvent. For values of $\chi < 1/2$, the polymer chains will expand into the solvent, for $\chi > 1/2$ the polymer chains attract and collapse. The steric stabilization thus will fail for less than θ-conditions. Solvents with $\chi < 1/2$ are called good solvents, solvents with $\chi > 1/2$ bad solvents.

For the sake of completeness, two mechanisms have to be mentioned. First, free polymer chains dissolved in the particle suspension can cause attraction between the particles. This effect is termed depletion interaction, as volumes depleted of polymer chains are formed around the particles. This volume is minimized by close contact among the particles and causes an effective attraction between the particles. Second, the two stabilization mechanisms, electrostatic by charges on the particle surfaces and steric by grafted polymer chains, can be combined to form an electrosteric stabilization. Both mechanisms are not exploited in this work.

Finally, by definition colloidal particles are suspended in a solvent. Dry structures are needed frequently for characterization and application of formed particle packing structures. This requires a transfer step with removal of the solvent by evaporation, draining or absorption. At some stage of the transfer step the particles will protrude through the solvent-air interface and experience capillary interactions. The attractive force F_i generated by supported particles deforming a liquid surface, often termed immersion force [19], can be approximated by

$$F_i = 2\pi\,\gamma_{lg}\,r_{lg}^2\,(\sin\Theta)^2\,\frac{1}{d}.$$ (1.11)

γ_{lg} is the surface tension of the solvent, Θ the contact angle of the solvent on the particle surface and r_{lg} is the radius of the three-phase contact line at the particle surface. With an $1/d$-dependency capillary interactions are long-ranged attractions, whose influence must be considered when investigating dried particle packings.

The plenty of interactions and mechanisms present may be summarized to "colloids are repulsive or attractive hard spheres performing random motion". The underlying principles of colloidal crystallization are easily elaborated in the following section using such a simplification.

1.3 Basic Principles of Colloidal Crystallization

A full overview of all experimental techniques to produce colloidal crystals is certainly beyond the scope of this introduction, and reviews are available [4, 5, 20–25]. However, using the simplification introduced in the previous section and basic thermodynamic ideas a classification of approaches to colloidal crystallization by self-assembly is given in the following.

Using the simple lattice model described by KEN A. DILL and SARINA BROMBERG [18], the basic crystallization models for spheres can be worked out. In this lattice model each of the N spheres occupies a site on a regular lattice and interacts with an energy of u_{NN} per bond only with particles present on the n_{NN} neighboring sites on a regular lattice. The total energy U of the particles on the lattice is easily calculated by the number of bonds m formed between neighboring particles,

$$U = m \cdot u_{NN}. \tag{1.12}$$

If the particles perform random thermal motion on the lattice, they experience translational entropy S. The free energy of the lattice model is then given by

$$F = U - TS. \tag{1.13}$$

The particles are free to move on the lattice and will occupy the state with the lowest free energy. Three implicit assumptions are made in this model. The spheres are assumed to be hard and cannot overlap and occupy the same lattice site. Additionally, their interactions are assumed isotropic with respect to the neighboring lattice sites. Finally, the range of the interactions between the particles must be smaller than their diameter so that only neighboring particles will interact.

In order to gain a classification of crystallization mechanisms we distinguish two extremes: $T = 0$ and $U = 0$. The first extreme is the case of granular matter, whose constituting objects do not perform random motion, but ballistic motion. The second extreme could be called an ideal gas or a hard-sphere-system, as the spheres on the lattice do not interact beside their hard bodies.

In the first extreme, $T = 0$, three crystallization mechanisms can be identified. The first mechanism occurs for $u_{NN} < 0$, i.e. attractive spheres. The minimal total energy is then given by

$$U_{min} = m \cdot u_{NN} = \frac{N n_{NN}}{2} \cdot u_{NN}, \tag{1.14}$$

a full occupation of the neighborhood of each sphere. This implies closest-packing and the formation of a crystal.

The opposite case of repulsive spheres, $u_{NN} > 0$, requires a full depletion of neighboring lattice sites for energy minimization:

$$U_{min} = m \cdot u_{NN} = 0 \cdot u_{NN} = 0. \tag{1.15}$$

Above a certain concentration of spheres on the lattice this condition can only be fulfilled by a regular or crystalline arrangement, any deviation would lead to a positive energy contribution. For example, in the simple case of a 3-dimensional cubic lattice a lattice site occupation of 50 % without energy penalty can only be realized by a perfect regular arrangement of the particles. This crystallization mechanism is very sensitive to the range of the interactions. With the repulsive interaction ranging further, more lattice sites would be unavailable to the spheres and they are forced to take a regular packing already at lower concentrations.

Finally, in case of no interaction among the particles besides the exclusion from occupied lattice sites, an externally applied potential can direct the packing of the spheres. The energy of each sphere u_i then becomes a function of a global distance parameter h, for example the height of the spheres in case of gravity as external potential. The total energy becomes the sum over the potential energies of the N particles. In the simple case of a monotonously attractive external potential the minimal total energy requires a minimal average $\overline{h}$:

$$U_{min} = \sum_{i=1}^{N} u_i(h) \Bigg|_{\overline{h} \to min} . \tag{1.16}$$

If the base area is fixed, a minimal $\overline{h}$ requires a minimal volume of the sphere packing and a maximal sphere concentration in the packing. The maximal sphere concentration occurs for the occupation of all lattice sites and the formation of a crystal. This captures both repulsive and attractive external potentials, because one can be transformed to another by inversion of the coordinates.

The second extreme, $U = 0$, is a fully entropy dominated regime. According to the equation of SACKUR-TETRODE [18], the entropy of an ideal gas is proportional to the natural logarithm of the volume per gas particle:

$$S = N k \ln \left(\frac{V_t}{N} \cdot T^{\frac{3}{2}} \right) + const. \tag{1.17}$$

The space filling of a packing defines the volume available to the spheres. In hexagonal close-packing 74 % of the total volume V_t is available to the spheres, in body-centered cubic packing 68 %, and in a dense random packing (BERNAL-glass) only 64 %. Regular arrangements thus provide the highest volume per sphere and, somewhat counterintuitive, maximizing the entropy requires an regular arrangement. A regular arranging of hard spheres with thermal motion is indeed predicted by simulations and found in experiments at 49 % volume concentration [26]. This entropy-driven crystallization is often termed WAINWRIGHT-ALDER-transition [27].

In summary, there are four principles of crystallization: minimization of the free energy

- by energy minimization under action of particle attraction,
- by energy minimization under action of particle repulsion,

- by energy minimization under action of an external potential,
- by entropy maximization.

This classification will be used in the following to categorize the present state of colloidal crystallization experiments.

1.4 Present State of Colloidal Crystallization

The classification given in the previous section is less clear-cut in physical systems. The basic assumptions made, $T = 0$ or $U = 0$, will not be realized in a real colloid, and entropic and energetic effects might both be present at the same time. For instance, colloidal particles may sediment and form a crystalline packing because of the external gravitation, or they may form a crystal because the settling increases the concentration at the bottom of the container until entropic crystallization sets in. This still may be distinguished by the interparticle distance in the crystal. If the particles formed a crystal by entropy maximization, they will try to keep a maximal interparticle spacing, while under the action of the gravitation alone the particle will pack with maximal density. In experiments, in which the volume available to the particles is continuously decreased until maximal density is reached, as in LANGMUIR-BLODGETT troughs or evaporating droplets, it is even less clear which mechanism originally induced formed crystallinity. The concentration increase in the suspension may lead to entropic ordering, to crystallization due to a possible repulsive potential and a reduction of the mean particle distance, or to crystallization by a possible attraction of the particles and a reduced entropic contribution to the free energy. As in most publications no statement on the underlying mechanism is made, the dominant contribution to the crystallization has to be assumed. For example, large particles with diameters of several hundred nanometers and low HAMAKER-coefficients like polymer latices and charge stabilization in an experiment with concentration increase are assumed to exhibit entropy dominated crystallization. Sterically stabilized metal nanoparticles in a precipitation experiment are assumed attraction dominated, and so forth.

By comparing the used colloid, the experimental procedure and the produced packing density, most experiments could be sorted to one of the crystallization mechanisms:

Crystallization by internal attraction. The dispersion interactions cause an omnipresent attraction among the particles. Increasing the dispersion interactions thus could be used to induce crystallization. However, it is cumbersome to increase the VAN DER WAALS-interactions, as it requires adjusting the permittivities of solvent and particles. Thus, typically either additional attractions are induced or the repulsion among the particles is reduced to gain an effective increase in attraction. Additional attractions were produced by addition of free polymers to induce depletion interactions [28–30], mixing of oppositely charge colloids [31–33], use of DNA-linkers [34] or photo-ionizable ligands [35] attached to the particles, or capillary interactions of particles in thin solvent films [36–39]. Repulsion was suppressed by lowering the solvent quality of sterically stabilized suspensions [4, 5, 40–42]. Adding a polar solvent to an unpolar suspension reduced χ until the steric stabilization failed. Suppression of charge-stabilization, e.g. by addition of ions to the suspension, usually leads to formation of low-density and fractal agglomerates rather than crystals [43–51].

Crystallization by internal attraction may not only be induced by effectively increasing the attraction, but also by reduction of the entropic contribution to the free energy of a moderately attractive colloid. The entropy can be reduced by reducing the volume per particle (compare Eq. 1.17). Reduction of the volume per particle is gained by evaporation of the solvent [52–62] or compressing floating particles in LANGMUIR-BLODGETT troughs [63–65].

The entropic contribution can also be reduced by lowering the temperature (Eq. 1.13). In case of steric stabilization lowering the temperature additionally reduces the repulsion (Eq. 1.9). Thus several effects potentially contribute when trying to induce crystallization by lowering the temperature. These effects are investigated in the second experimental part of this thesis.

Crystallization by internal repulsion. The range of the repulsive interactions has to be increased for crystallization by internal repulsion. Charge-stabilized suspensions are ideally suited for this purpose. The range of the interactions is defined by the DEBYE screening length (Eq. 1.9), which can be adjusted by the concentration of ions or the pH of the solvent. Removal of ions by osmosis or resins increases the range of the repulsions, which eventually induces crystallization [66–71].

Crystallization in an external potential. An easy way to produce colloidal crystals is sedimentation of the particles [72–74]. This method is restricted to large colloidal particles, which exhibit significant settling. Other approaches include electrodynamic, electrostatic and magnetic potentials [75–77], or assembly in solvent flow fields [78–97].

The latter approach represents the interesting case of a non-conservative force field associated with a potential-like energy of the particles. This is investigated in the first experimental part of this work.

Crystallization by entropy maximization. Entropy maximization is a common way to produce colloidal crystals. It requires well-stabilized suspensions where the range of the particle repulsion is small compared to the particle diameter and renders the particles hard-sphere like. The suspensions can be directly prepared with particle concentrations suitable for crystallization [30, 98–100]. Alternatively, the particle concentration is continuously increased by evaporation until crystallization is gained, often inside of emulsion droplets [101–105].

The variety of ready-to-use concepts for crystallization brings about the question of remaining frontiers in colloidal crystallization. Beyond the present limitations of the applicability of colloidal crystals mentioned in the motivation, several issues arise from deviations from the assumptions underlying the discussion presented in this chapter. The colloidal particles are assumed spherical, hard, possessing isotropic short interactions, and performing ballistic motion superimposed with random thermal motion. Numerous theoretical publications focus on colloids deviating from these assumptions. The minimal free-energy packings of non-spherical particles are investigated by simulations [106, 107]. The phase behavior of soft colloids are investigated using the example of star-polymers [108–110]. Non-isotropic interactions are common situation in protein suspensions [111] or model patchy particles [112, 113]. Thermodynamic phase diagrams and colloidal crystallization are predicted to strongly depend on the range of the interparticle interactions [114, 115]. Finally the assumption of pure ballistic and diffusive motion might break down for colloids in close contact. Contact friction, an effect known from granular matter, may become significant [116].

In contrast to the simple schemes of crystal formation, no general description of the mechanisms preventing crystallization is available. Attempts to solve this problem are made using phase diagrams in analogy to jamming phase diagrams known from granular matter [117, 118]. Factors generally assumed to hinder formation of crystals are a high polydispersity of the particles [119–121] and a competing glass transition of the particles [9, 122].

1.5 Choice of Experimental Systems

The aim of this work is to find principles for the formation of high-quality colloidal crystals from nanoparticles with control of structure formation beyond hexagonal close-packing.

Notwithstanding the mentioned ambiguities, the presented classification of crystallization methods aids selecting appropriate approaches and development of new approaches. The crystallization by internal repulsion of the particles and the entropic crystallization lead to a low density packing of the particles, while the crystallization by attraction among the particles and by confinement in an external potential leads to close-packed structures. For the production of materials by self-assembly the latter two approaches thus are advantageous. They offer to form mechanically stable structures which can by dried without altering the structure. The crystals with maximal spacing between the particles require an additional fixation of the packing structure to be extracted from the suspension.

The assembly in solvent flow fields has the least restrictions and requirements in terms of particle material, particle size, substrate and setup compared to the other crystallization approaches using external potentials. Particle assembly in convective fluid flows is potentially continuous, fast, and can be designed to directly deposit the particles on substrates. This makes this approach useful for applications. However, this approach does not yet provide the production of large-area high quality particle packings needed for application. The quality of the particle packings is limited by small-grained polycrystallinity, thickness transitions, holes and drying cracks. Additionally, there are uncertainties on the underlying assembly mechanisms, and the approach is only established for large particles with diameters of several hundred nanometers. Consequently, improving the quality of the formed particle packings and clarify the assembly mechanisms are the areas where further research appears necessary. In this work a deposition setup is constructed which allows investigation of the arrangement process of individual particles and of the liquid meniscus in which the assembly takes place. As model suspension charge-stabilized emulsion-polymerized polystyrene particles are used. They are available with very low size dispersities over a large range of sizes. In Chap. 2 relations between meniscus geometry and particle film quality are developed and results of interference microscopy measurements on the meniscus geometry are presented. In Chap. 3 the arrangement process is investigated *in-situ* down to a single-particle level using particle image velocimetry and particle tracking to develop a detailed understanding of the crystallization mechanism and to predict parameters for the production of crystals from nanoscaled colloids.

Crystallization by internal attraction has been proven feasible by several approaches. However, thermal induction and control of crystallization in sterically stabilized suspensions has not been tested yet. Thermal induction and control of crystallization is seemingly simple, as it does not involve foreign solvents like in precipitation experiments, moving phase boundaries like in solvent evaporation driven experiments, or even altering the particles to bear linkers like DNA-molecules on the surface. It also promises to be a robust approach, as lowering the temperature reduces the solvent quality (Eq. 1.9), reduces the repulsive potential generated by the ligand chains (Eq. 1.10), and reduces the entropic contribution counteracting the energetic contribution to the free energy (Eq. 1.13). Sufficiently lowering the temperature thus should always lead to a minimal energy packing of the particles, independent of particle size and material. Analogies of this cooling approach to colloidal crystallization to the condensation of molecular or atomic gases by cooling including formation of kinetically controlled morphologies of crystalline packings may be expected (compare Fig. 1.1). Consequently, comprehensive experiments on such a new approach including assembly kinetics, structure formation, and crystallization mechanisms appear necessary. Gold nanoparticles with alkyl thiol ligands in unpolar solvents are used as model systems. These particles can be synthesized with sub-10 nm diameter with low size dispersities. Their high HAMAKER-coefficient and stabilization with short alkyl-chains promise to provide interparticle attractions with practical transition temperatures. The much smaller size of the particles compared to the polymer particles used in the first experimental part of this work requires a different set of experimental techniques. In Chap. 4 the kinetics of the agglomeration process are investigated with dynamic light scattering and compared with results on the formed particle packings gained by small-angle X-ray scattering and transmission electron microscopy. In Chap. 5 crystallization is induced by combining temperature and foreign solvent to steer the packing. In addition to electron microscopy and X-ray scattering, differential scanning calorimetry and diffusing wave spectroscopy are used to evaluate mechanism hindering colloidal crystallization.

The underlying question as stated in Chap. 1 is common in both approaches: under which external conditions do colloidal crystals form and how can the packing and morphology be controlled beyond the particle shape? The apparent disparities between the two used approaches and model suspensions enables additionally to look for common principles enforcing or hindering colloidal crystallization and for structure formation processes by self-assembly beyond hexagonal hard sphere packing, spanning two orders of magnitude.

This thesis is composed of four parts: this introductory part, two experimental parts and a final concluding part. In this introductory part motivation for the experiments performed is given. Terms used throughout the work are defined, basic interactions and mechanisms present in colloidal suspensions and leading to crystals are discussed and, finally, the experiments are rationalized. The experimental parts are divided into two chapters. The first chapters each introduce a route towards colloidal crystals and the possibility to produce structured materials is investigated. In the second chapters the underlying crystallization mechanisms are investigated in detail. In the concluding part the insights from the different experiments are summarized and an outlook on future experiments is given.

References

1. E. Roduner, *Nanoscopic Materials Size-Dependent Phenomena* (RSC Publishing, Cambridge, 2006)
2. V. Rotello, *Nanoparticles Building Blocks for Nanotechnology*. (Springer, New York, 2004)
3. D.J. Norris, A view of the future. Nat. Mater. **6**, 177–178 (2007)
4. C.B. Murray, C.R. Kagan, M.G. Bawendi, Synthesis and characterization of monodisperse nanocrystals and close-packed nanocrystal assemblies. Annu. Rev. Mater. Sci. **30**, 545–610 (2000)
5. D.V. Talapin, J.-S. Lee, M.V. Kovalenko, E. Shevchenko, Prospects of colloidal nanocrystals for electronic and optoelectronic applications. Chem. Rev. **110**, 389–458 (2010)
6. W. Bentley, File:Snowflake4.png. Wikimedia Commons (1902). http://commons.wikimedia.org
7. W.B. Russel, D.A. Saville, W.R. Schowalter, *Colloidal Dispersions* (Cambridge University Press, Cambridge, 1989)
8. J. Lyklema, *Fundamentals of Interface and Colloid Science. Fundamentals*, vol. 1, (Academic Press, New York, 1991)
9. P.N. Pusey, in *Colloidal Suspensions*, ed. by J.P. Hansen, D. Levesque, J. Zinn-Justin. Liquids freezing and glass transition (North-Holland, Amsterdam, 1991), pp. 763–942
10. D.F. Evans, H. Wennerström, *The Colloidal Domain: Where Physics, Chemistry, Biology and Technology Meet* (Wiley-VCH, New York, 1999)
11. H.M. Jaeger, S.R. Nagel, R.P. Behringer, The physics of granular materials. Phys. Today **49**, 32–38 (1996)
12. C. Kittel, *Einführung in die Festkörperphysik* (Oldenbourg, Munich, 1999)
13. G.M. Whitesides, M. Boncheva, Beyond molecules: self-assembly of mesoscopic and macroscopic components. Proc. Nat. Acad. Sci. **99**, 4769–4774 (2002)
14. G.M. Whitesides, B. Grzybowski, Self-assembly at all scales. Science **295**, 2418–2421 (2002)
15. J. Israelachvili, *Intermolecular and Surface Forces* (Academic Press, London, 2011)
16. V.A. Parsegian, *Van der Waals forces* (Cambridge University Press, New York, 2006)
17. A.K. Doland, S.F. Edwards, Theory of the stabilization of colloids by adsorbed polymer. Proc. Royal Soc. London A Math. Phy. Sci. **337**, 509–516 (1974)
18. K.A. Dill, S. Bromberg, *Molcular Driving Forces: Statistical Thermodynamics in Chemistry and Biology* (Garland Science, Taylor & Francis, 2002)
19. P.A. Kralchevsky, K. Nagayama, *Particles at Fluid Interfaces and Membranes* (Elsevier, Amsterdam, 2001)

20. A. Yethiraj, Tunable colloids: control of colloidal phase transitions with tunable interactions. Soft Matter **3**, 1099–1115 (2007)
21. Y. Xia, B. Gates, Y. Yin, Y. Lu, Monodispersed colloidal spheres: old material with new applications. Adv. Mater. **12**, 693–712 (2000)
22. F. Li, D.P. Josephson, A. Stein, Kolloidale Organisation: der Weg vom Partikel zu kolloidalen Molekülen und Kristallen. Angewandte Chemie **123**, 378–409 (2011)
23. S. Kinge, M. Cego-Calama, D.N. Reinhoudt, Self-assembling nanoparticles at surfaces and interfaces. ChemPhysChem **9**, 20–42 (2008)
24. F. Marlow, Muldarisnur, P. Sharifi, R. Brinkmann, C. Mendive, Opals: statis and prospects. Angewandte Chemie (International Edition) **48**, 6212–6233 (2009)
25. B.L.V. Prasad, C.M. Sorensen, K.J. Klabunde, Gold nanoparticle superlattices. Chem. Soc. Rev. **37**, 1871–1883 (2008)
26. P.N. Pusey, E. Zaccarelli, C. Valeriani, E. Sanz, W.C.K. Poon, M.E. Cates, Hard spheres: crystallization and glass formation. Philos. Trans. R. Soc. A **367**, 4993–5011 (2009)
27. B.J. Alder, T.E. Wainwright, Phase transition for a hard sphere system. J. Chem. Phys. **27**, 1208–1209 (1957)
28. F. Renth, W.C.K. Poon, R.M.L. Evans, Phase transition kinetics in colloid-polymer mixtures at triple coexistence: kinetic maps from free-energy landscapes. Phys. Rev. E **64**, 031402-1-031402-9(2001)
29. V.J. Anderson, H.N.W. Lekkerkerker, Insights into phase transition kinetics from colloid science. Nature **416**, 811–815 (2002)
30. A.P. Gast, W.B. Russel, Simple ordering in complex fluids. Phys. Today **1998**, 24–30 (1998)
31. M.E. Leunissen, C.G. Christova, A.-P. Hynninen, C.P. Royall, A.I. Campbell, A. Imhof, M. Dijkstra, R. van Roij, A. van Blaaderen, Ionic colloidal crystals of oppositely charged particles. Nature **437**, 235–240 (2005)
32. E.C.M. Vermolen, A. Kujik, L.C. Filion, M. Hermes, J.H.J. Thijssen, M. Dijkstra, A. van Blaaderen, Fabrication of large binary colloidal crystals with NaCl structure. Proc. Nat. Acad. Sci. **106**, 16063–16067 (2009)
33. A.M. Kalsin, M. Fialkowski, M. Paszewski, S.K. Smoukov, K.J. Bishop, B.A. Grzybowski, Electrostatic self-assembly of binary nanoparticle crystals with diamond-like lattice. Science **312**, 420–424 (2006)
34. D. Nykypanchuk, M.M. Maye, D. van der Lellie, O. Gang, DNA-guided crystallization of colloidal nanoparticles. Nature **451**, 549–552 (2008)
35. R. Klajn, K.J. Bishop, B.A. Grzybowski, Light-controlled self-assembly of reversible and irreversible nanoparticle superstructures. Proc. Nat. Acad. Sci. **104**, 10305–10309 (2007)
36. N.D. Denkov, O.D. Velev, P.A. Kralchevsky, I.B. Ivanov, H. Yoshimura, K. Nagayama, Mechnism of formation of two-dimensional crystals from latex particles on substrates. Langmuir **8**, 3183–3190 (1992)
37. N. Aubry, P. Singh, M. Janjua, S. Nudurupati, Micro- and Nanoparticles self-assembly for virtually defect-free, adjustable monolayers. Proc. Nat. Acad. Sci. **105**, 3711–3714 (2008)
38. V. Santhanam, J. Liu, R. Agarwal, R.P. Andres, Self-assembly of uniform monolayer arrays of nanoparticles. Langmuir **19**, 7881–7887 (2003)
39. L. Cui, Y. Zhang, J. Wang, Y. Ren, Y. Song, L. Jang, Ultra-fast fabrication of colloidal photonic crystals by spray coating. Macromol. Rapid Commun. **30**, 598–603 (2009)
40. D.V. Talapin, E.V. Shevchenko, A. Kornowski, N. Gaponik, M. Haase, A.L. Rogach, H. Weller, A new approach to crystallization of CdSe nanoparticles into ordered three-dimensional super-lattices. Adv. Mater. **13**, 1868–1871 (2001)
41. D.V. Talapin, E.V. Shevchenko, C.B. Murray, A. Kornowski, S. Förster, H. Weller, CdSe and CdSe/CdS nanorod solids. J. Am. Chem. Soc. **126**, 12984–12988 (2004)
42. M.B. Sigman, A.E. Saunders, B.A. Korgel, Metal nanocrystal superlattice nucleation and growth. Langmuir **20**, 978–983 (2004)
43. D.A. Weitz, J.X. Zhu, D.J. Durian, H. Gang, D.J. Pine, Diffusing-wave spectroscopy: the technique and some applications. Phys. Scr. **T49**, 610–621 (1993)

44. M.Y. Lin, H.M. Lindsay, D.A. Weitz, R.C. Ball, R. Klein, P. Meakin, Universal reaction-limited colloidal aggragation. Phys. Rev. A **41**, 2005–2020 (1990)
45. M.Y. Lin, H.M. Lindsay, D.A. Weitz, R. Klein, R.C. Ball, P. Meakin, Universal diffusion-limited colloidal aggragation. J. Phys.: Condens. Matter **2**, 3093–3113 (1990)
46. M.Y. Lin, H.M. Lindsay, D.A. Weitz, R.C. Ball, R. Klein, P. Meakin, Universality in colloidal aggragation. Nature **339**, 360–362 (1989)
47. P. Dimon, S.K. Sinha, D.A. Weitz, C.R. Safinya, G.S. Smith, W.A. Varady, H.M. Lindsay, Structure of aggregated gold colloids. Phys. Rev. Lett. **57**, 595–598 (1986)
48. D.A. Weitz, M.Y. Lin, Dynamic scaling of cluster-mass distribution in kinetic colloid aggregation. Phys. Rev. Lett. **57**, 2037–2040 (1986)
49. D.A. Weitz, M.Y. Lin, C.J. Sandroff, Colloidal aggregation revisited: new insights based on fractal structure and surface-enhanced Raman scattering. Surf. Sci. **158**, 147–164 (1985)
50. D.A. Weitz, J.S. Huang, M.Y. Lin, J. Sung, Limits of the fractal dimension for irreversible kinetic aggregation of gold colloids. Phys. Rev. Lett. **54**, 1416–1419 (1985)
51. D.A. Weitz, J.S. Huang, M.Y. Lin, J. Sung, Dynamics of diffusion-limited kinetic aggregation. Phys. Rev. Lett. **53**, 1657–1660 (1984)
52. B.A. Korgel, S. Fullam, S. Connolly, D. Fitzmaurice, Assembly and self-organization of silver nanocrystal superlattices: ordered "Soft Spheres". J. Phys. Chem. B **102**, 8379–8388 (1998)
53. B.A. Korgel, D. Fitzmaurice, Small-angle x-ray-scattering study of silver-nanocrystal disorder-order phase transitions. Phys. Rev. B **59**, 14191–14201 (1999)
54. B. Korgel, D. Fitzmaurice, Condensation of ordered nanocrystal thin films. Phys. Rev. Lett. **80**, 3531–3534 (1998)
55. C. Gutierrez-Wing, P. Santiago, J. Ascencio, A. Camacho, M. Jose-Yacaman, Self-assembling of gold nanoparticles in one, two, and three dimensions. Appl. Phys. A **71**, 237–243 (2000)
56. A. Dong, X. Ye, J. Chen, C.B. Murray, Two-dimensional binary and ternary nanocrystal superlattices: the case of monolayers and bilayers. Nano Lett. **11**, 1804–1809 (2011)
57. E.V. Shevchenko, D.V. Talapin, N.A. Kotov, S. O'Brien, C.B. Murray, Structural diversity in binary nanoparticle superlattices. Nature **439**, 55–59 (2006)
58. S. Disch, E. Wetterskog, R.P. Hermann, G. Salazar-Alvarez, P. Busch, T. Bruckel, L. Bergstrom, S. Kamali, Shape induced symmetry in self-assembled mesocrystals of iron oxide nanocubes. Nano Lett. **11**, 1651–1656 (2011)
59. E. Klecha, D. Ingert, M.P. Pileni, How the level of ordering of 2D nanocrystal superlattices is controlled by their deposition mode. J. Phys. Chem. **1**, 1616–1622 (2010)
60. N. Goubet, J. Richardi, P.A. Albouy, M.P. Pileni, How to predict the growth mechanism of supracrystals from gold nanocrystals. J. Phys. Chem. Lett. **2**, 417–422 (2011)
61. R.L. Whetten, J.T. Khoury, M.M. Alvarez, S. Murthy, I. Vezmar, Z.L. Wang, P.W. Stephens, C.L. Cleveland, W.D. Luedkte, U. Landmann, Nanocrystal gold molecules. Adv. Mater. **8**, 428–433 (1996)
62. S.A. Harfenist, Z.L. Wang, R.L. Whetten, I. Vezmar, M.M. Alvarez, Three-dimensional hexagonal close-packed superlattice of passivated Ag nanocrystals. Adv. Mater. **9**, 817–822 (1997)
63. J.R. Heath, C.M. Knobler, D.V. Leff, Pressure/temperature phase diagrams and superlattices of organically functionalized metal nanocrystal monolayers: the influence of particle size, size distribution, and surface passivant. J. Phys. Chem. B **101**, 189–197 (1997)
64. I.L. Pei, K. Mori, M. Adachi, Investigation on arrangement and fusion behaviors of gold nanoparticles at the air/water interface. Colloids Surf. A **281**, 44–50 (2006)
65. A. Tao, P. Sinsermsuksakul, P. Yang, Tunable plasmonic lattices of silver nanocrystals. Nature **2**, 435–440 (2007)
66. P. Wette, I. Klassen, D. Holland-Moritz, D.M. Herlach, H.J. Schöpe, N. Lorenz, H. Reiber, T. Palberg, S.V. Roth, Complete description of re-entrant phase behavior in a charge variable colloidal model system. J. Chem. Phys. **132**, 131102-1–131102-4 (2010)
67. T. Palberg, Crystallization kinetics of repulsive colloidal spheres. J. Phys.: Condens. Matter **11**, R323–R360 (1999)
68. D.J.W. Aastuen, N.A. Clark, L.K. Cotter, B.J. Ackerson, Nucleation and growth of colloidal crystals. Phys. Rev. Lett. **57**, 1733–1736 (1986)

69. A. Kose, M. Ozakia, K. Takanoa, Y. Kobayashia, S. Hachisu, Direct observation of ordered latex suspension by metallurgical microscope. J. Colloid Interface Sci. **44**, 330–338 (1973)

70. R. Goldberg, H.J. Schöpe, Opaline pydrogels: polycrystalline body-centered-cubic bulk material with an in-situ variable latice constant. Chem. Mater. **19**, 6095–6100 (2007)

71. D.M. Herlach, I. Klassen, P. Wette, D. Holland-Moritz, Colloids as model systems for metal alloys: a case study of crystallization. J. Phys.: Condens. Matter **22**, 153101-1-153101-20 (2010)

72. O. Vickreva, O. Kalinina, E. Kumacheva, Colloid crystal growth under oscillatory shear. Adv. Mater. **12**, 110–112 (2000)

73. P. Schall, I. Cohen, D.A. Weitz, F. Spaepen, Visualization of dislocation dynamics in colloidal crystals. Science **305**, 1944–1948 (2004)

74. P. Schall, I. Cohen, D.A. Weitz, F. Spaepen, Visualization of dislocation nucleation by indenting colloidal Crystals. Nature **440**, 319–323 (2006)

75. A. von Blaaderen, Colloids under external control. MRS Bull. **2004**, 85–90 (2004)

76. S. Kim, R. Asmatulu, H.L. Marcus, F. Papadimitrakopoulos, Dielectrophoretic assembly of grain-boundary-free 2D colloidal single crystals. J. Colloid Interface Sci. **354**, 448–454 (2011)

77. C. Zhu, L. Chen, H. Xu, Z. Gu, A magnetically tunable colloidal crystal film for reflective display. Macromol. Rapid Commun. **30**, 1945–1949 (2009)

78. T.P. Bigioni, X.-M. Lin, T.T. Nguyen, E.I. Corwin, T.A. Witten, H.M. Jaeger, Kinetically driven self assembly of highly ordered nanoparticlemonolayers. Nat. Mater. **5**, 265–270 (2006)

79. C.D. Dushkin, G.S. Lazarov, S.N. Kotsev, H. Yoshimura, K. Nagayama, Effect of growth conditions on the structure of two-dimensional latex crystals: experiment. Colloid Polym. Sci. **277**, 914–930 (1999)

80. A.D. Dimitrov, A. Nagayama, Continuous convective assembling of fine particles into two-dimensional arrays on solid surfaces. Langmuir **12**, 1303–1311 (1996)

81. N.D. Denkov, O.D. Velev, P.A. Kralchevsky, I.B. Ivanov, H. Yoshimura, K. Nagayama, Two-dimensional crystallization. Nature **361**, 26 (1993)

82. D.J. Norris, E.G. Arlinghaus, L. Meng, R. Heiny, L.E. Scriven, Opaline photonic crystals: How does self-assembly work? Adv. Mater. **16**, 1393–1399 (2004)

83. L. Meng, H. Wei, A. Nagel, B.J. Wiley, L.E. Scriven, D.J. Norris, The role of thickness transitions in convective assembly. Nano Lett. **6**, 2249–2253 (2006)

84. E. Vekris, V. Kitaev, D.D. Perovic, J.S. Aitchinson, G.A. Ozin, Visualization of stacking faults and their formation in colloidal photonic crystal films. Adv. Mater. **20**, 1110–1116 (2008)

85. K. Chen, S.V. Stoianov, J. Bangerter, H.D. Robinson, Restricted meniscus convective self-assembly. J. Colloid Interface Sci. **344**, 315–320 (2010)

86. J. Kleinert, S. Kim, O.D. Velev, Electric-field-assisted convective assembly of colloidal crystal coatings. Langmuir **26**, 10380–10385 (2010)

87. B.G. Prevo, D.M. Kuncicky, O.D. Velev, Engineered deposition of coatings from nano- and micro-particles: a brief review of convective assembly at high volume fraction. Colloids Surf. A: Physicochem. Eng. Aspects **311**, 2–10 (2007)

88. X. Checoury, S. Enoch, C. Lopez, A. Blanco, Stacking patterns in self-assembly opal photonic crystals. Appl. Phys. Lett. **90**, 161131 (2007)

89. P. Jiang, J.F. Bertone, K.S. Hwang, V.L. Colvin, Single-crystal colloidal multilayers of controlled thickness. Chem. Mater. **11**, 2132–2140 (1999)

90. P. Kumnorkaew, Y.-K. Ee, N. Tansu, J.F. Gilchrist, Investigation of the deposition of microsphere monolayers for fabrication of microlens arrays. Langmuir **24**, 12150–12157 (2008)

91. L. Malaquin, T. Kraus, H. Schmid, E. Delamarche, H. Wolf, Controlled particle placement through convective and capillary assembly. Langmuir **23**, 11513–11521 (2007)

92. G.S. Lozano, L.A. Dorado, R.A. Depine, H. Miguez, Towards a full understanding of the growth dynamics and optical response of self-assembled photonic colloidal crystal films. J. Mater. Chem. **19**, 185–190 (2009)

93. H. Fan, K. Yang, D.M. Boye, T. Sigmon, K.J. Malloy, H. Xu, G.P. Lopez, C.J. Brinker, Self-Assembly of ordered, robust, three-dimension gold nanocrystal/silica arrays. Science **304**, 567–571 (2004)

94. H.J. Schöpe, A.B. Fontecha, H. König, J.M. Hueso, R. Biehl, Fast microscopic method for large scale determination of structure, morphology, and quality of thin colloidal crystals. Langmuir **22**, 1828–1838 (2006)

95. T. Still, W. Cheng, M. Retsch, U. Jonas, G. Fytas, Colloidal systems: a promising material class for tailoring sound propagation at high frequencies. J. Phys.: Condens. Matter **20**, 404203-1-404203-9 (2008)

96. A.G. Marin, H. Gelderblom, D. Lohse, J.H. Snoeijer, Order-to-disorder transition in ring-shaped colloidal stains. Phys. Rev. Lett. **107**, 085502-1–085502-4 (2011)

97. E.C.H. Ng, K.M. Chin, C.C. Wong, Controlling inplane orientation of a monolayer colloidal crystal by meniscus pinning. Langmuir **27**, 2244–2249 (2011)

98. P.N. Pusey, W. van Megen, Phase behavior of concentrated suspensions of nearly hard colloidal spheres. Nature **320**, 340–342 (1986)

99. K.E. Davis, W.B. Russel, W. Glantschning, Disorder-to-order transition in settling suspensions of colloidal silica: X-ray measurments. Science **306**, 507–510 (1989)

100. J. Zhu, M. Lin, R. Rogers, W. Meyer, R.H. Ottewill, S.-.S.S. Crew, W.B. Russel, P.M. Chaikin, Crystallization of hard-sphere colloids in microgrvity. Nature **387**, 883–885 (1997)

101. Y.-S. Cho, S.-H. Kim, G.-R. Yi, S.-M. Yang, Self-organization of colloidal nanospheres inside emulsion droplets: high-order clusters, supraparticles, and supraballs. Colloids Surf. A **345**, 237–245 (2009)

102. M.I. Bodnarchuk, L. Li, A. Fok, S. Nachtergaele, R.F. Ismagilov, D.V. Talapin, Three-dimensional nanocrystal superlattices grown in nanoliter microfluidic plugs. J. Am. Chem. Soc. **133**, 8956–8960 (2011)

103. O.D. Velev, K. Furusawa, K. Nagayama, Assembly of Latex particles by using emulsion droplets as templates. 1. Microstructured hollow spheres. Langmuir **12**, 2374–2384 (1996)

104. O.D. Velev, K. Furusawa, K. Nagayama, Assembly of Latex particles by using emulsion droplets as templates. 2. Ball-like and composite aggregates. Langmuir **12**, 2385–2391 (1996)

105. O.D. Velev, A.M. Lenhoff, E.-W. Kaler, A class of microstructural particles through colloidal crystallization. Science **287**, 2240–2243 (2000)

106. Z. Zhang, S.C. Glotzer, Self-assembly of Patchy particles. Nano Lett. **4**, 1407–1413 (2004)

107. A.W. Wilber, J.P.K. Doye, A.A. Louis, E.G. Noya, M.A. Miller, P. Wong, Reversible self-assembly of patchy particles into monodisperse icosahedral clusters. J. Chem. Phys. **127**, 085106-1– 085106-11 (2007)

108. K.A. Dawson, The glass paradigm for colloidal glasses, gels, and another arrested states driven by attractive interactions. Curr. Opin. Colloid Interface Sci. **7**, 218–227 (2002)

109. G. Foffi, G.D. McCullagh, A. Lawlor, E. Zaccarelli, K.A. Dawson, F. Sciortino, P. Tartaglia, D. Pini, G. Stell, Phase equilibria and glass transition in colloidal systems with short-ranged attractive interactions: application to protein crystallization. Phys. Rev. E **65**, 031407-1–031407-17 (2002)

110. S.V. Karpov, I.L. Isaev, A.P. Gavrilyuk, V.S. Gerasimov, A.S. Grachev, General principles of the crystallization of nanostructured disperse systems. Colloid J. **71**, 313–328 (2009)

111. V. Trappe, V. Prasad, L. Cipelletti, P.N. Segre, D.A. Weitz, Jamming phase diagram for attractive particles. Nature **411**, 772–775 (2001)

112. A.J. Liu, S.R. Nagel, Jamming is not just cool any more. Nature **396**, 21–22 (1998)

113. M. Fasolo, P. Sollich, Equilibrium Phase behavior of polydisperse hard spheres. Phys. Rev. Lett. **91**, 068301-1–068301-4 (2003)

114. S.-E. Phan, W.B. Russel, J. Zhu, P.M. Chaikin, Effects of polydispersity on hard sphere crystals. J. Chem. Phys. **108**, 9789–9795 (1998)

115. N. Geerts, S. Jahn, E. Eiser, Direct observation of size fractionation during colloidal crystallization. J. Phys.: Condens. Matter **22**, 104111-1–104111-5 (2010)

116. K.J.M. Bishop, C.E. Wilmer, S. Soh, B.A. Grzybowski, Nanoscale forces and their uses in self-assembly. Small **5**, 1600–1630 (2009)

117. S.C. Glotzer, M.J. Solomon, Anisotropy of building blocks and their assembly into complex structures. Nat. Mater. **6**, 557–562 (2007)

118. Z.Z.M.A. Horsch, M.H. Lamm, S.C. Glotzer, Thethered nano building blocks: toward a conceptual framework for nanoparticle self-assembly. Nano Lett. **3**, 1341–1346 (2003)
119. M. Watzlawek, C.N. Likos, H. Löwen, Phase diagram of star polymer solutions. Phys. Rev. Lett. **82**, 5289–5292 (1999)
120. A. Jusufi, J. Dzubiella, C.N. Likos, C. von Ferber, H. Löwen, Effective interactions between star polymers and colloidal particles. J. Phys.: Condens. Matter **13**, 6177–6194 (2001)
121. C.N. Likos, Effective interactions in soft condensed matter. Phys. Rep. **348**, 267–439 (2001)
122. J.P.K. Doye, A.A. Louis, C. Lin, L.R. Allen, E.G. Noya, A.W. Wilber, H.C. Koke, R. Lyus, Controlling crystallization and its absence: proteins, colloids and patchy models. Phys. Chem. Chem. Phys. **9**, 2197–2205 (2007)

Part I
Convective Particle Assembly

Chapter 2
Role of Meniscus Shape in Large-Area Convective Particle Assembly

Abstract Dense and uniform particle films are deposited using a robust version of the convective particle assembly process. We analyze how the shape of the gas-liquid interface and the three-phase contact line govern the stability of convective deposition and thus, the achievable quality of films. Interference microscopy indicates that a highly curved meniscus cannot compensate for the ubiquitous perturbation during deposition. A moderately curved meniscus provides flexibility to compensate and localize perturbation and enables reliable homogeneous deposition. We analyze which setup geometry and meniscus velocity yield appropriate meniscus shapes. The quality of the resulting films is analyzed and compared to the deposition conditions. Uniform films over areas beyond the centimeter range are accessible using the optimized process, which is suitable for functional particle coatings and templates for microstructured materials.

2.1 Introduction

Convective particle assembly [1] yields dense films of microparticles and nanoparticles. The process has found applications in the deposition of functional materials, patterning of surfaces and many other fields. The reliable deposition of homogeneous particle films over macroscopic areas beyond millimeters remains challenging, however. Further research is necessary to make convective assembly sufficiently robust and reliable for the many applications that require high-quality particle films:

Optical coatings, which include plasmonic structures [2], photonic bandgap materials [3–6] and microlens arrays on top of light emitting diodes [7] manipulate light depending on structure periodicity. Deviations from this periodicity drastically change the reflection and transmission properties—defects scatter light, locally increased particle distances or distortion change optical bandgaps or disturb the alignment with underlying structures. Other applications require constant thickness without voids to guarantee a constant, low dielectric constant so that the layers can be used as a low-k interlayer materials [8], protect an underlying surface, as in antire-

P. G. Born, *Crystallization of Nanoscaled Colloids*, Springer Theses, DOI: 10.1007/978-3-319-00230-9_2, © Springer International Publishing Switzerland 2013

flective and self-cleaning coatings on solar cells [9] or enhance analytical methods, as in layers for surface enhanced Raman spectroscopy [10, 11] and nanocavity arrays for surface assisted mass spectroscopy [12]. Finally, when particle layers are used as masks for nanosphere lithography [13], defects propagate to the final structured material.

Research on convective particle assembly has focused on the application of convectively assembled films as optical active material [3–7, 9, 14], on the characterization of the microscopic structure of the films [3, 5, 7, 15–20] and on the role of solvent flow in the assembly process [2, 21–24]. A limited number of studies focus on the aspect of setup geometry and meniscus shape [7].

The simplest setup for the convective assembly of particles is an evaporating suspension droplet on a solid substrate [25], but more sophisticated setups have been developed to provide precise control over the deposition parameters and assemble particle films on large areas. These setups can be divided into three principal types:

First, setups with static substrates immersed vertically or at an angle into an evaporating suspension bath (Fig. 2.1a) [3, 4, 14, 18, 19, 21–23]. This type of setup is easily realized and allows the production of bulk colloidal crystals. It requires larger sample volumes and longer production times than other setups and provides limited control over the film thickness because the particle volume fraction progressively increases during deposition.

The second type of setup resembles conventional dip coating where a substrate is withdrawn vertically from a suspension reservoir (Fig. 2.1b) [5, 6, 17, 20, 26, 27]. In this geometry particle films can be deposited on substrates of almost arbitrary sizes within minutes or hours. The film thickness is set by the withdrawal velocity. Observation of the process is possible in situ using a horizontally mounted long working distance microscope objective. Drawbacks again include the large sample volume needed and limited access for in situ observation.

The third type of setup replaces the reservoir bath by a small amount of suspension held by capillary forces between the substrate and a blade (Fig. 2.1c). This small reservoir is moved over the entire substrate by moving the blade and the substrate against each other [2, 7, 11, 24, 28, 29]. Deposition rates similar to those in dip-coating can be reached, with several advantages: The required sample volume is minimized to tens of microliters. The substrate size is unlimited. The suspension reservoir forms a capillary bridge independent of gravity that can be tilted arbitrarily, which facilitates integration into light microscopes. In-situ observation of particle assembly via microscopy allows precise control over the film thickness. The setup is also ideally suited to experimentally explore deposition parameters. For example, temperature control of the substrate is straightforward via the back side of the substrate [24], the compact configuration is readily integrated into small chambers with controlled air flow over the evaporating area [2] and AC fields can be applied between blade and substrate to induce electrowetting [29].

Common to all described configurations is the convective assembly process that takes place in a meniscus at the edge of a suspension reservoir. The geometry of the meniscus resembles a curved wedge between the bulk reservoir and a microscopically thin solvent film on the wetted substrate. Pronounced evaporation from this liquid

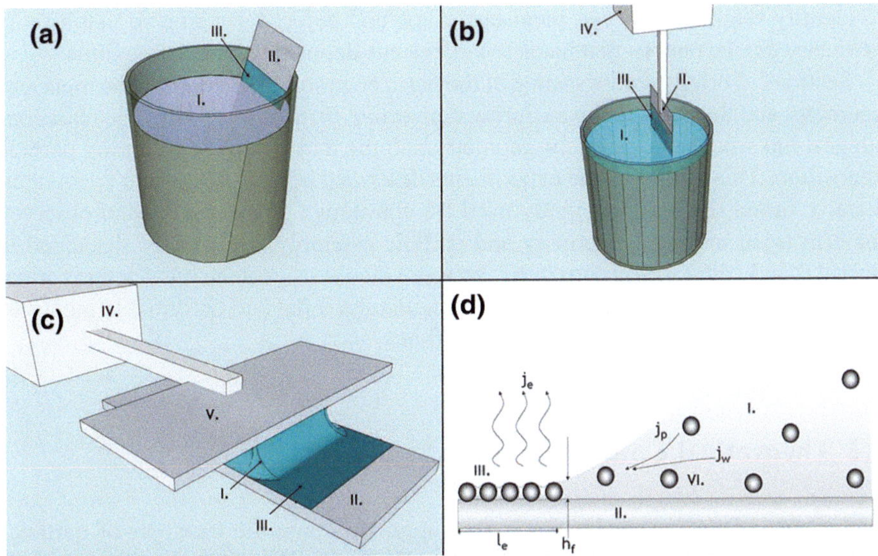

Fig. 2.1 The different types of deposition setups for convective particle assembly (**a–c**) and the common underlying process (**d**). **a** Deposition on an immersed static substrate (*II.*): the film (*III.*) forms as liquid evaporates from a reservoir (*I.*), gradually lowering its surface. **b** Deposition in dip-coating geometry: a motor (*IV.*) pulls a substrate (*II.*) from a reservoir (*I.*). **c** Blade-coating-geometry: the reservoir (*I.*) is held between a substrate (*II.*) and a blade (*V.*) which is pulled by a motor (*IV.*). **d** The microscopic process is identical in all setups: pronounced evaporation from the wet particle film (*III.*) induces a convective flux (*VI.*) from the bulk reservoir (*I.*) into the film. The suspended particles are dragged along and are wedged into a densely packed film. The symbols j_e, j_w and j_p denote the evaporative, the water and the particle flux, respectively, h_f indicates the height of the film and l_e the length of wet film from which evaporation occurs

film induces a convective flow from the bulk liquid into the thin film that transports and confines the particles (Fig. 2.1d). The meniscus height limits the film growth in the surface-normal direction.

While the assembly principle is identical for all setups, the shape of the gas-liquid interface is markedly different. The substrate either rests or moves with respect to the reservoir so that the meniscus exhibits either its static shape or a dynamic shape depending on the relative velocity. The meniscus is either constrained or unconstrained, depending on whether the setup imposes (for example, by a blade) a certain shape on it or the meniscus can freely adapt to the substrate. The substrate is either fully immersed into the suspension or the suspension rest on the substrate and forms a long contact line on it. So far, the influences of the meniscus shape on the formation of defects in the deposited films have rarely been addressed in research. Macroscopic defects of the deposited films are caused by contact line depinning, local growth rate variations and variations in the thickness of the grown film. These mechanisms depend on contact line dynamics, the particle flux and the space available for assembly, all of which are connected to the meniscus shape. In this study, we aim to identify relations between meniscus shape and defect formation so that a setup geometry can be chosen that leads to a meniscus depositing defect-free films.

Section 2.2 briefly reviews some of the basic equations that describe the meniscus geometry and the particle fluxes during convective particle assembly. The equations suggest links between meniscus geometry and the defect formation during particle deposition. This motivates the experiments described in Sect. 2.3, where we systematically varied the setup geometry used for convective self-assembly and observed the effects on meniscus geometry and particle assembly. The results, discussed in Sect. 2.4, indicate a strong correlation between meniscus geometry and defect formation processes as predicted by theory. In particular, a flat and deformable meniscus proves highly beneficial for robust deposition.

2.2 Theoretical Considerations

Convective particle assembly is, as first approximation, the transport of particles from a bulk suspension into a dense film. It takes place inside a meniscus on a wetting substrate that has a shape and properties which we will discuss in Sect. 2.2.1. The equations that describe the transport of liquid and particles during assembly are discussed in Sect. 2.2.2. In Sect. 2.2.3, relations between assembly and meniscus geometry are established.

In the following, we will use three parameters to characterize the meniscus shape: The curvature of the contact line on the substrate $1/R_{\parallel}$, the curvature of the meniscus surface perpendicular to the substrate $1/R_{\perp}$ and the contact angle of the meniscus, Θ (Fig. 2.2). We analyze how they depend on the constraints that the setup imposes and how they affect the quality of assembled films.

2.2.1 The Shape of a Meniscus on a Wetting Substrate

General shape. The general shape of a static liquid meniscus on a wetting substrate has been evaluated by Sujanani and Wayner using interference microscopy [30]. They describe three regions: First, a flat adsorbed film with a thickness on the order of tens of nanometers which is mainly governed by disjoining pressure and the dynamics of condensation and evaporation. There follows a highly curved segment which connects the thin film region to a region dominated by capillary forces. This transition segment features the highest curvature of the meniscus, at a liquid thickness of about 100 nm. The meniscus ends with the central region that is dominated by capillary forces and exhibits a low, constant curvature $1/R_{\perp}$. The conventional, macroscopic contact line is obtained by extending the constant-curvature segment to the substrate so that it forms a macroscopic contact angle Θ. In this work, we focus on regions of the meniscus that are thicker than 100 nm. We assume that the general description of the meniscus shape holds true for this part of the meniscus

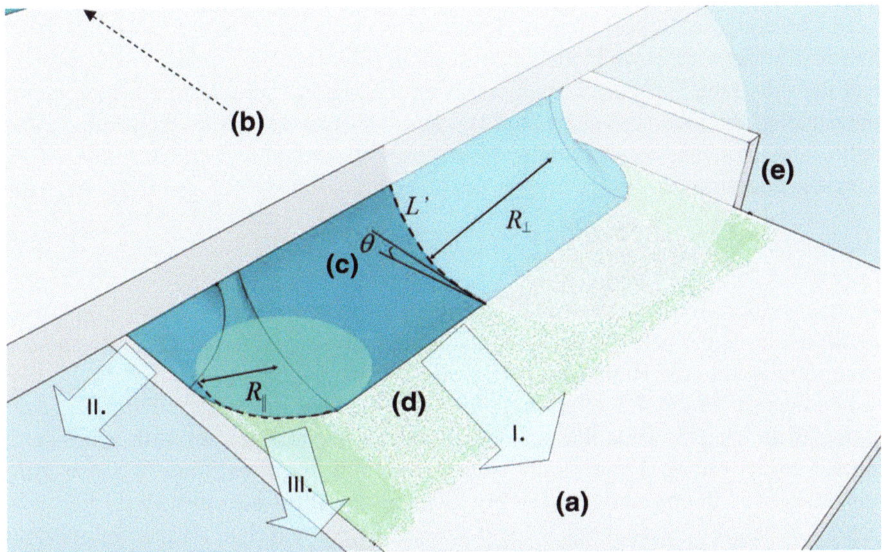

Fig. 2.2 The meniscus geometry in convective particle assembly between a substrate **a** and a blade **b**. The suspension reservoir **c** leaves a wet particle film **d** as it moves between barriers **e** that prevent evaporation flux to the sides (*arrow II.*). Evaporation occurs at the linear segment of the contact line (*arrow I.*) and at the curved segment of the contact line (*arrow III.*). The geometry is characterized by the radius of curvature of the contact line on the substrate $R_\parallel$, the radius of curvature of the meniscus perpendicular to the substrate $R_\perp$ and the contact angle of the meniscus on the substrate Θ. The crossection of the meniscus' surface between blade and substrate is a circular segment of length L'. The dotted arrow indicates the relative motion of the blade with respect to the substrate

during particle deposition. Exceptions where the dynamic meniscus is pinned during certain perturbations of the deposition are discussed below.

Contact angle. The contact angle Θ formed by the static meniscus in contact with the substrate is described by the law of YOUNG- LAPLACE [31],

$$\gamma_{lg} \cos(\Theta_s) = \gamma_{sg} - \gamma_{sl}, \tag{2.1}$$

where γ_{lg} is the liquid-gas interfacial energy, γ_{sg} the solid-gas interfacial energy, γ_{sl} the solid-liquid interfacial energy and Θ_s the static contact angle. As indicated by Eq. 2.1, the equilibrium value of the contact angle is sensitive to changes in the surface energies. Temperature changes affect the surface energies, as estimated by EÖTVÖS' law [32]

$$\gamma = \gamma_0 - \gamma_T \cdot T, \tag{2.2}$$

where γ_0 and γ_T are material parameters and T the absolute temperature. This law implies a linear decline in surface energy with increasing temperature. As all involved surface energies are altered, the change of the contact angle with temperature cannot be generally derived using above equations and has to be determined experimentally for the involved substrate and solvent pairs.

In addition, the contact angle is sensitive to the velocity of the substrate. When the substrate is withdrawn from the suspension reservoir, the contact line is dragged over the substrate. Viscous dissipation inside the meniscus hinders an instant reconfiguration of the meniscus shape and the liquid surface assumes a dynamic contact angle. This dynamic contact angle Θ_d has been described by DE GENNES [33–35] using the equation

$$\Theta_d(\Theta_s^2 - \Theta_d^2) = \frac{\eta}{\gamma_{lg}} \cdot 6 \ln \frac{S}{a} \cdot v_{cl}, \qquad (2.3)$$

where η is the viscosity of the solvent, S is a typical length scale or size of the meniscus, v_{cl} is the velocity of the contact line with respect to the substrate and a is a length related to the Hamaker coefficient A, $a \equiv \sqrt{A/6\pi\gamma_{lg}}$.

Application of Eq. 2.3 requires some caution. First, slippage effects had been neglected in Eq. 2.3. Thus the equation is valid only for menisci with small static contact angle, $\Theta_s \ll 1$ rad $\approx 57°$, an assumption that is fulfilled by the wetting substrates used in convective assembly. Second, the substrate velocity v_{cl} has to be small compared to a maximal velocity v_m of the contact line. At this velocity the difference between Θ_s and Θ_d is maximal, the contact line vanishes and a continuous film is left on the substrate by the receding contact line. For water on a substrate with a static contact angle of $2°$ the minimal dynamic contact angle and the maximum velocity can be estimated using Eq. 2.3 to $\Theta_d \approx 1.15°$ and $v_m \approx 200\ \mu\text{m/s}$ [33]. This maximal velocity of the contact line also sets a maximal velocity for convective particle assembly, because the contact line velocity must be matched by the substrate withdrawal velocity, as discussed below. Third, Eq. 2.3 again describes the macroscopic contact angle determined by the central uniformly curved part of the meniscus.

In the boundaries set by these assumptions, Eq. 2.3 predicts a decrease of the contact angle with v_{cl}. Thus, in setups with moving substrates, the deposition generally takes place in menisci with lower contact angles than in setups with static substrates, where the meniscus assumes the static contact angle.

Meniscus curvature. The meniscus intersects with the substrate at the contact line, where it assumes a defined contact angle Θ. In a dip-coating geometry, the transition from the reservoir's liquid surface to the substrate surface imposes a curvature on the liquid surface. The curvature of the surface is determined by the liquid surface tension. For unconstrained menisci the radius of curvature $R_\perp$ is on the order of the capillary length $\kappa^{-1} = \sqrt{\gamma_{lg}/\rho_l g}$ of the solvent (where ρ_l is the density of the liquid and g the gravitational acceleration) [33].

In a confined geometry, the blade sets boundary conditions that can only be fulfilled by a curved meniscus. In setups where the reservoir is held by capillary interactions between a blade and a substrate, the distance between blade and substrate H has to be smaller than the capillary length κ^{-1}. At distances H larger than the capillary length, the anchoring of the reservoir between the blade and the substrate would fail. The curvature of the reservoir is a function of the lower and upper contact angles and the distance H between blade and substrate. For an ideal capillary bridge between two wetting substrates, the curvature scales with the inverse distance H

between the two anchoring plates [33],

$$\frac{1}{R_\perp} \propto \frac{1}{H} \geq \frac{1}{\kappa^{-1}}. \tag{2.4}$$

Thus, the curvature $1/R_\perp$ is larger for setups with a blade than in setups without a blade.

Contact line curvature. LAPLACE's equation applies for all points on the meniscus surface if it is in equilibrium with a reservoir: $1/R_1 + 1/R_2 = \Delta p/\gamma_{lg} = const.$ Pressure differences Δp may arise from hydrostatic pressure inside the meniscus. We have shown above that the meniscus in convective particle assembly is curved with $1/R_\perp$. Thus, the meniscus must exhibit a second principle curvature with opposite sign to compensate. The meniscus is curved not only about an axis that is parallel to the substrate with a radius $R_\perp$ but also about an axis normal to the substrate with a radius $R_\parallel$. Since the contact line is the intersection of the meniscus with the substrate, it follows this curvature. Since both curvatures compensate for each other, the curvature of the contact line reflects the setup geometry in the same way as the meniscus curvature does.

Meniscus deformability. Above relations describe the influence of substrate velocity and blade height on the meniscus geometry, namely, meniscus curvature and contact angle. Meniscus geometry not only sets the space that is available for particle motion, it also governs the meniscus' deformability: The force required to locally deform the meniscus depends on its overall shape. De Gennes and coworkers developed a model for the deformability by distortions that can be described with low wavenumber harmonics [33, 34, 36, 37]. They found distorted and undistorted solutions for LAPLACE's condition of vanishing total curvature and used the area added by the distortion to calculate a restoring force. This force can be expressed in terms of the line shape $\xi(d)$ that a defect of size d imposes on the contact line of the meniscus (Fig. 2.3) and a "spring constant" k of the contact line,

$$k = \frac{\pi \gamma_{lg} \Theta^2}{\ln\frac{L}{d}}. \tag{2.5}$$

The constant L describes the macroscopic distance between adjacent anchor points of the contact line. The energy U_{def} added by the larger area of the distorted meniscus is given by:

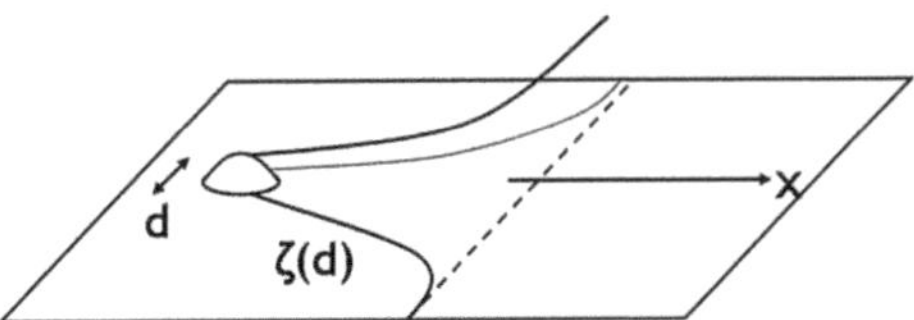

Fig. 2.3 Deformation of the contact line by a surface inhomogeneity of size d. The undisturbed meniscus moves over the inhomogeneity in x-direction and gets pinned. The contact line deforms to the new shape $\xi(d)$

$$U_{def} = \frac{1}{2} \cdot k \cdot \xi^2(d). \tag{2.6}$$

Note the dependence of the spring constant on the contact angle Θ and on $\ln(L/d)$. A meniscus with a small contact angle and a long distance L between the anchor points of the contact line will exhibit a small spring constant and a high deformability. One might call such a meniscus "slack" and a meniscus with large contact angle and small distance between anchor points "taut".

DE GENNES et al. derive the spring constant considering only anchor points along the contact line. However, their derivation is general enough to describe a meniscus that is pinned not only on the substrate, but also perpendicular to it as occurs in deposition setups with blades. The height of the blade above the substrate is generally much smaller than the width of the substrate and even smaller than the capillary length. The relevant distance to the next anchor point that restricts the deformation of the meniscus L is then given by the distance L' along the meniscus surface from the defect to the blade. If we can approximate this path by a circular segment with radius $R_\perp$ between substrate and blade (Fig. 2.2), the relevant distance calculates to

$$L' \propto 2\pi \cdot R_\perp. \tag{2.7}$$

A meniscus with large $R_\perp$ will thus be slack, while a highly curved meniscus with low $R_\perp$ is taut. Both $R_\perp$ and contact angle Θ depend on setup geometry and assembly velocity. Setups without constraints (with weakly curved menisci) and assembly at high velocities (at low dynamic contact angles) provide slack menisci. Constraints or low velocities provide taut menisci.

DE GENNES' model explains an additional effect of the blade. When L' is the distance between the contact line and the blade's edge, measured along the meniscus surface, Eq. 2.5 predicts a diverging spring constant k when the deformation size d reaches the order of the characteristic distance L. In setups with blades, we can assume $L = L'$. The blade therefore restricts the curvature $1/R_\parallel$ and forces a straight contact line. In contrast, the meniscus is curved along the full contact line for unconstrained menisci.

2.2.2 Growth Rate of Convectively Assembled Particle Films

The shape of the meniscus sets the space that is available for the particles to move and assemble into a film. Their net motion has been described by DIMITROV and NAGAYAMA in 1995 [26]. In their work, they balance the liquid flows and connect them with the particle fluxes to obtain simple transport equations. The evaporative flow of the solvent, $J_e = w_f \cdot l_e \cdot j_e$, and the solvent flow from the bulk suspension, $J_w = w_f \cdot l_f \cdot j_w$, must balance in steady state:

$$l_e j_e = l_f j_w, \tag{2.8}$$

where w_f, l_e, h_f, and j denote the width of the deposited film, the length of wet film measured from the deposition front, the height of the film and the solvent flux densities, respectively (Fig. 2.1). The particle flux density j_p is then linked to the solvent flux density by a coupling parameter, β, and the particle volume fraction in the liquid, ϕ:

$$j_p = \frac{\beta\phi}{1-\phi} j_w.$$ (2.9)

The particle flux to the film growth front, $J_p = w_f \cdot h_f \cdot j_p$, matches the increase in particle volume of the film, which is the product of the film growth rate v_f, the film thickness h_f, the film width w_f and the film density $(1 - \epsilon)$ (where ϵ is the porosity), $J_{vol} = w_f \cdot h_f \cdot v_f \cdot (1 - \epsilon)$, so that we obtain a connection between growth rate and particle flux:

$$v_f(1 - \epsilon) = j_p.$$ (2.10)

Combining Eqs. 2.8, 2.9 and 2.10 yield an equation for the film growth rate:

$$v_f = \frac{l_e \beta j_e \phi}{h_f(1 - \epsilon)(1 - \phi)}$$ (2.11)

If the meniscus does not move, these transport conditions only lead to an accumulation of particles at the contact line. For a continuous deposition of a film with constant thickness, the contact line must move at the same rate as the film grows: $v_{cl} = v_f$. The motion of the contact line can either result from an actively moved substrate or from an evaporating reservoir on a static substrate. Note that relation 2.11 is to some extent self-equilibrating. For a given velocity v_{cl}, the particle film thickness h_f increases or decreases until equilibrium is reached.

2.2.3 Relations Between Film Growth and Meniscus Shape

The balance of fluxes that yields Eq. 2.11 does not fully describe the assembly process. Evaporation length, evaporative flux and deposited film thickness are variables that depend on the geometry of the meniscus in which assembly takes place. Perturbations of the assembly process lead to unbalanced fluxes that are not covered by the theory. In the following, we will discuss how the meniscus geometry sets some of the variables in Eqs. 2.8, 2.9, 2.10 and 2.11. We also analyze the effects of perturbations, which in fact are dependent on the meniscus shape.

Thickness variations and meniscus slope. The thickness of the deposited film h_f is not a continuous value but an integer multiple of the thickness of a single particle layer. The film thickness depends on the height of the meniscus above the growth front, h_m, which therefore equals m times the thickness of a single particle layer plus a fraction of a particle diameter (Fig. 2.4). If the front of the growing particle film

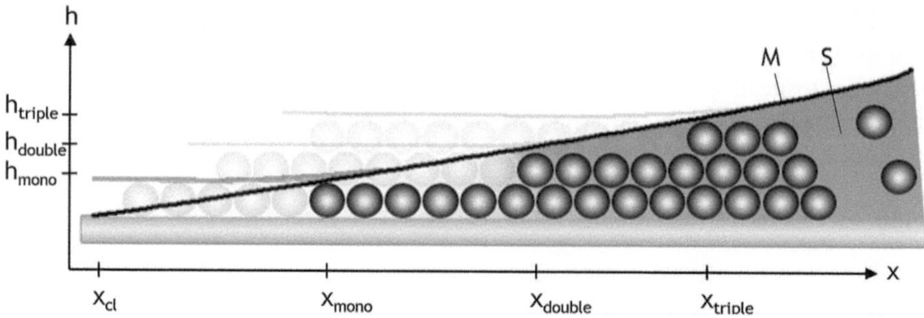

Fig. 2.4 The relation between the thickness of the deposited film and the meniscus height. The suspension (S) forms a meniscus with a defined gas-liquid interface (M) terminated by a contact line on the moving substrate at x_{cl}. The convective fluxes lead to an accumulation of particles in the meniscus and initiate the growth of a particle monolayer at x_{mono}, where the height of the meniscus h_m equals the particle diameter. If the monolayer grows faster than the substrate moves, the film front moves into positive x direction, towards the bulk suspension. When it reaches x_{double}, where h_m equals the height of a particle double layer h_{double}, the assembly of a double layer begins

is at an x-position of the meniscus that can accommodate a maximum of m particle layers, a film of just this thickness will grow. The distance between x-positions that can accommodate different m depends on the shape of the meniscus. Its height h_m grows from the contact line towards the reservoir depending on the contact angle Θ at the edge and on the meniscus' curvature normal to the substrate $1/R_\perp$.

If the meniscus is steep, the deposition will show a high tendency to switch between film thicknesses if the assembly fronts moves inside the meniscus. The deposition then has a low "buffer length" l_{buffer} that can compensate for mismatches $\Delta v = |v_{cl}| - |v_f|$ between contact line velocity and growth velocity. This sets a "buffer time" $t_{buffer} = l_{buffer}/\Delta v$, which is the time between the onset of a mismatch and a change in deposited layer thickness (with all its consequences). Large buffer lengths prevent local variations in particle or solvent flux from changing the layer geometry and let random variations cancel out over time. Long buffer times also give the operator more time to readjust parameters such as the substrate velocity to minimize Δv. A large buffer length requires a low contact angle Θ and a low curvature $1/R_\perp$. We have previously seen that a high withdrawal speed and a large substrate-blade distance provide both.

Growth rate and contact line curvature. The evaporative flux density j_e drives the assembly process. Strong evaporation causes a large water flux, a large particle flux and a high film growth rate. The evaporative flux depends on two ambient parameters, temperature and humidity, and on the local curvature of the liquid surface. DEEGAN and coworker described the relation between the curvature of a spherical droplet of radius R and the evaporation rate by treating the purely diffusive evaporative flux as an electrostatic potential emerging from a charged conductor [25, 38, 39]. Diffusion-limited evaporation leads to a steady-state concentration profile $u(r)$ of vapor in air so that the diffusion equation is reduced to a Laplace equation, $D\Delta u = \partial_t u = 0$, where D is the diffusion constant for the vapor in air and u the mass of vapor per unit volume of air. The boundary conditions impose saturation of the air above the liquid surface at $u(0)$ and ambient vapor concentration $u(\infty)$ at a large distance from the drop, a boundary value problem identical to that of a charged conductor. The evaporative flux density from the surface of a sessile drop with a contact line that is curved at a radius $R_\parallel$ thus depends on the distance r from the axis of rotational symmetry according to

$$je \propto \frac{1}{(R_\parallel - r)^\lambda}. \tag{2.12}$$

where $\lambda = (\pi - 2\Theta)/(2\pi - 2\Theta)$ takes the contact angle into account. For any point at a constant distance r from the axis, the evaporative flux will increase as the radius of curvature decreases. It diverges in the wedge near the contact line at $r \approx R_\parallel$.

If the particle film growth front is situated at a constant r, the radius of curvature of the associated contact line segment determines the evaporation rate. Comparing this relation between contact line curvature $1/R_\parallel$ and the evaporative flux density je to the equation describing the growth rate of the film (Eq. 2.11), we expect an increased growth rate at curved segments of the contact line.

The length of the contact line affects deposition, too. If the contact line is curved, its full length exceeds its length projected onto the film growth front. This surplus in the actual length compared to the length projected onto the film growth front tends to increase deposition. The surplus is given by a factor $\sin \alpha$, if α is the angle between contact line and growth direction.

Local differences in contact line geometry will therefore influence the local growth rate. In any real (finite) setup, the contact line is curved, so that the deposition will be inhomogeneous over the full substrate width, an effect that cannot be compensated by the choice of withdrawal velocity. It is possible to avoid contact line curvature at the initial stage of deposition using structured substrates [40]. However, the contact line will relax to its equilibrium shape as the meniscus moves away from the surface relief. It is therefore desirable to have a setup where the contact line exhibits little curvature over most of the substrate. This is the case for setups with blades that pin the meniscus, force a straight appearance of the contact line perpendicular to the growth direction and thus deposit more homogeneously.

Evaporation length and meniscus deformability. Holes form in the deposited layer when the thin liquid film ruptures and the contact line jumps to a new position without depositing particles. Holes perturb the deposition process because the evaporation length l_e drops to zero behind them (Fig. 2.5a). A vanishing evaporation length leads to a minimal film growth velocity v_f (Eq. 2.11) and continuously induces new holes in the film (Fig. 2.5b). An additional buffer that helps preventing this defect propagation is set by the "depinning length". During normal deposition, the contact line is pinned at the film's growth front. If the film's growth velocity is insufficient and the growth front moves in negative x direction, away from the equilibrium position of the contact line, the contact line will be dragged along up to a maximum displacement that we call depinning length l_{dep}. The time that passes before the contact line detaches from the growth front and a hole forms is $t_{dep} = l_{dep}/\Delta v$. Similar to the buffer length, a long depinning length is beneficial because it increases the stability against

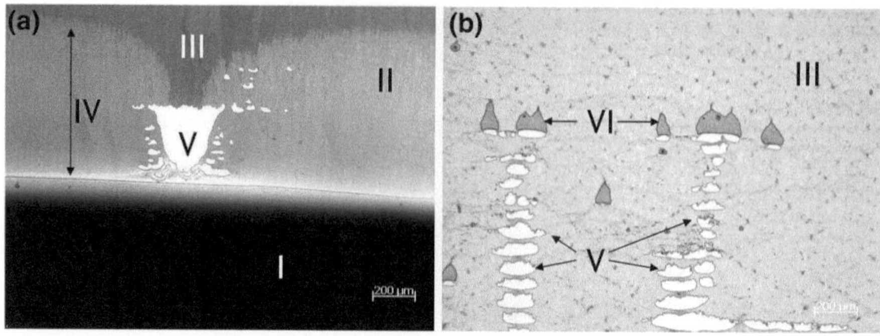

Fig. 2.5 Light microscopy images of the deposition process (**a**) and a dried film (**b**) of a 500 nm polystyrene particle film with defects. The meniscus of the suspension (*I*) guides particles into a wet particle film (*II*), which dries at a distance from the meniscus (*III*) that defines the evaporation length (*IV*). The evaporation length l_e decreases when holes (*V*) form in the film, for example behind double layers (*VI*). This induces the nucleation of new holes and leads to a film with chains of holes (**b**)

perturbations. If the growth rate v_f fluctuates during deposition, for example due to an increased evaporation length l_e or due to changes in the particle volume fraction ϕ, it needs to be constantly readjusted by the operator. A slack meniscus with large l_{dep} gives the operator additional time to readjust the system. Moderate fluctuations cancel out without intervention.

A slack meniscus also limits the size of defects. Both contact line curvature and local defects (dust particles, holes or the local deposition of additional particle layers) violate the balance of Eq. 2.11. To maintain continuous deposition despite of such perturbations, the meniscus must adapt locally. If the meniscus is sufficiently deformable, defects due to the violated balance will remain localized. Figure 2.6 shows how a taut meniscus propagates defects over large distances, while a slack meniscus adapts and localizes the holes. Both depinning length and deformability depend on the spring constant k of the meniscus. Low spring constants are achieved at high withdrawal speeds and for large blade-substrate distance.

Evaporation at substrate sides. Substrates pin the suspension at their edges. At the pinning edges, menisci form, strong evaporation occurs and particles accumulate. This leads to enhanced deposition of particles at the substrate edges. For setups with substrates fully immersed into the suspension this is a marginal problem, but for setups with the suspension confined under a blade this leads to significant loss of particles and unwanted deposits at the sides of the substrate (Fig. 2.7a). Diffusion barriers that prevent evaporation to the sides (Fig. 2.2) provide an effective remedy (Fig. 2.7b).

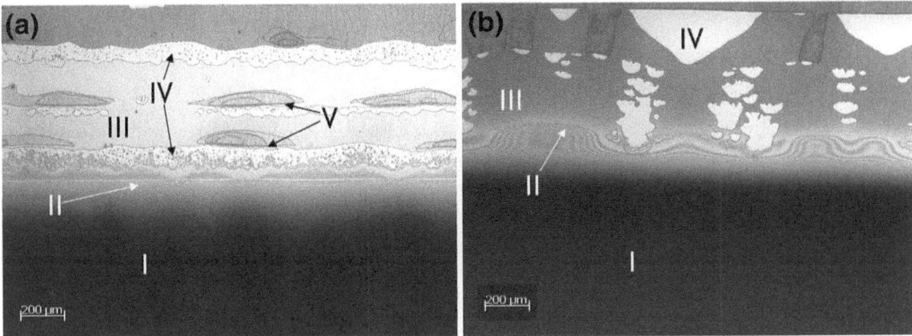

Fig. 2.6 Interference microscopy shows the deposition of films from suspensions of 500 nm polystyrene particles under different process conditions. The meniscus (*I*) is terminated by a contact line (*II*) which moves toward the bottom of the images and deposits particle films (*III*). Common defects in the films are gaps (*IV*) and multilayer (*V*). In **a**, the deposition takes place at 1 μm/s withdrawal velocity and 160 μm blade height. The meniscus is straight and stiff and the contact line detaches over large segments from the particle film. In **b**, deposition takes place at 10 μm/s withdrawal velocity and 1 mm blade height. The meniscus is deformable and the contact line stays pinned at the film front so that defects remain localized

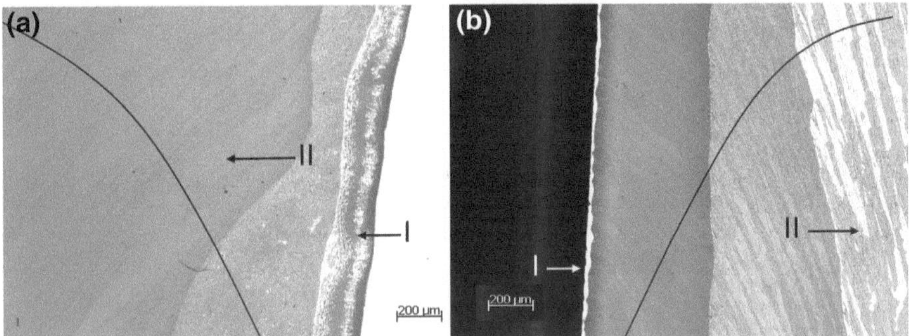

Fig. 2.7 Light microscopy images showing the edge of dry particle films deposited without **a** and with **b** a barrier to prevent evaporation to the sides. The dark lines indicate the shapes of the contact lines during deposition. Region *I*, directly at the edge, shows a pileup of particles (**a**), which is minimized by introducing the barrier (**b**). Region *II* shows multilayer that were deposited from the curved segment of the contact line and could not be prevented by the evaporation barriers

2.3 Results

We used a setup for convective self-assembly from a restricted meniscus inside a light microscope to capture the geometry of the meniscus during assembly. Height information was obtained via interference light microscopy illuminated by a monochromatic light emitting diode (Fig. 2.16). Reflection light microscopy yielded the shape of the projection of the meniscus on the substrate and thus, the shape of the contact line. Interference light microscopy allowed precise measurement of the height of the meniscus and calculation of contact angle and curvature. In Sects. 2.3.1–2.3.4, we show how contact line curvature, contact angle, meniscus curvature and meniscus deformability depend on blade height and substrate withdrawal velocity. The results guide our choice of blade heights and velocities for deposition experiments in Sect. 2.3.5, where we evaluate the effects of geometry on the convective deposition of particles. Finally, we present parameters that yield homogeneous films from 500 nm particles.

2.3.1 Effect of Blade Height on Contact Line Curvature

The static, particle-free meniscus exhibits several properties that are important during assembly. In our experimental setup, the shape of the contact line was fixed by three edges, two at the sides of the substrate and one at the front of the blade (Fig. 2.2). The pinning at the front edge of the blade forced the contact line on the substrate into a virtually straight line parallel to the blade that was terminated by two curved transition segments towards the edges of the substrate. The width of the curved segments depended on the setup geometry and grew with increasing blade height H as measured by light microscopy (Fig. 2.8a). When increasing the blade height from 20 μm to 1 mm, the width of the curved segments increased from roughly half a millimeter to 3.5 mm each, shortening the straight contact line from both sides. This increase is due to the larger L' at higher blade height H which reduce the effect of the blade. At the same time, the curvature $1/R_\parallel$ of the curved segments of the contact line dropped to one quarter of its original magnitude. This decrease implies a reduced curvature $1/R_\perp$ of the meniscus.

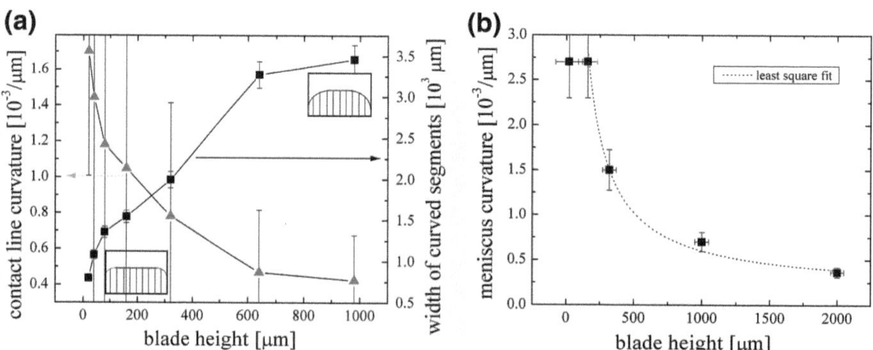

Fig. 2.8 Shape of the static meniscus depending on the blade-substrate-distance as measured by interference light microscopy. Graph **a** shows the curvature $1/R_\parallel$ and extension of the curved segments of the contact line on the substrate. Graph **b** shows the curvature $1/R_\perp$ of the meniscus perpendicular to the substrate. Lines between the points in **a** are guidelines for the eyes only, lines between the points in **b** are a least-squares fit of Eq. 2.4, neglecting the leftmost point

Increasing the blade height lowered the curvature and reduced the difference in evaporation rates between the curved segments and the straight segments of the contact line, thereby reducing differences in film growth rates. This improvement in deposition uniformity comes at the cost of straight contact line length.

2.3.2 Effect of Blade Height and Substrate Velocity on Meniscus Curvature

The curvature of the meniscus $1/R_\perp$ that is normal to the contact line curvature also depends on the blade height (Fig. 2.8b). We find a good fit when describing the relationship using the equation

$$\frac{1}{R_\perp} = \frac{1}{2 \cdot H}.$$

(2.13)

derived from Eq. 2.4 with a proportionality factor of two. The values for the two curvatures $1/R_\parallel$ and $1/R_\perp$ match within the experimental errors, both drop from $\sim 2 \cdot 10^{-3}$ to $\sim 4 \cdot 10^{-4}$ $\mu\mathrm{m}^{-1}$ with raising the blade height.

Film deposition requires a moving meniscus. The dynamic shape of a particle-free meniscus captures further aspects of the deposition process. We reconstructed the shape of a water meniscus while the substrate was withdrawn and determined its curvature as a function of velocity (Fig. 2.9). For low velocities, the curvature $1/R_\perp$ did not change with withdrawal velocity. This is consistent with the geometrical considerations in Sect. 2.2.1. For velocities above 50 μm/s, the curvature decreased. At such velocities, viscous effects gain influence. The segment of the meniscus under investigation takes an intermediate shape between the static shape and the limiting dynamic case predicted by Eq. 2.3, where a flat, continuous film is extracted from the liquid meniscus at very high velocities.

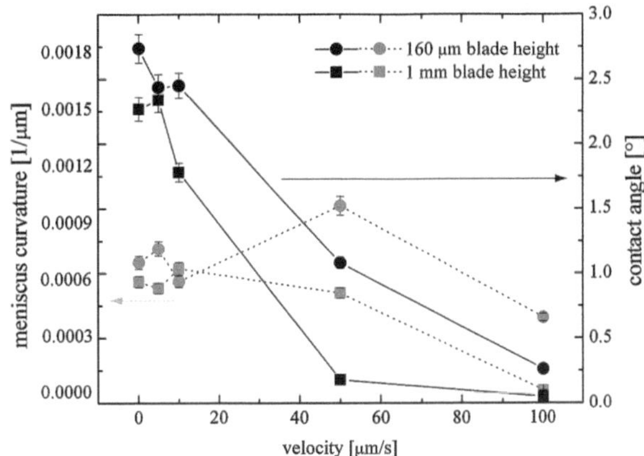

Fig. 2.9 Shape of the dynamic meniscus on a withdrawn substrate as measured by interference light microscopy. The curvature $1/R_\perp$ of the meniscus perpendicular to the substrate and its contact angle Θ with the substrate were measured as functions of the withdrawal velocity for two blade heights. The lines are meant to guide the eyes

2.3.3 Effect of Substrate Velocity on Contact Angle

The response of the contact angle to changing substrate velocity is shown in Fig. 2.9. The dynamic contact angle drops by two orders of magnitude in the measured velocity range below 100 μm/s, more than expected from Eq. 2.3. The experimental contact angle is calculated from the interference fringes, which are measured down to heights below 100 nm. These heights are at the border of the central segment, for which Eq. 2.3 is valid, to the proximal highly curved segment of the meniscus. The proximal segment thus exhibits much stronger deformation than the central segment during the transition from the static contact angle to the continuous film extraction. This effect can influence the convective assembly process at least of small particles.

2.3.4 Effects of Blade Height and Substrate Velocity on Meniscus Deformability

We investigated the effect of changing curvature and contact angle on disturbances of the meniscus by measuring the depinning length l_{dep} of the contact line at an artificial defect. A microscopic epoxy obstacle was placed on the substrate and moved from the solvent reservoir through the meniscus at different blade heights and substrate velocities, an experiment similar to that of NADKARNI and GAROFF [41]. The distance between the obstacle's edge and the undisturbed contact line was measured at the moment of depinning (Fig. 2.10) to obtain l_{dep}. Figure 2.11 shows an increase of the depinning length with withdrawal speed, in particular at large blade heights. The results match the expectations from Eq. 2.5. The lower dynamic contact angle Θ_d and the larger distance to the anchor points L' at high withdrawal velocities and large blade distances led to a smaller spring constant and a greater depinning length.

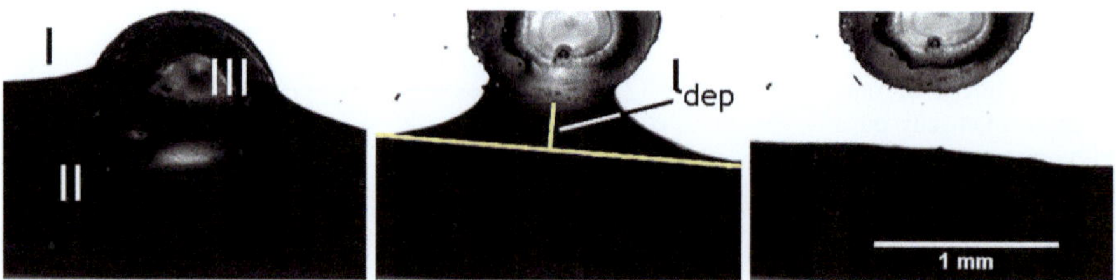

Fig. 2.10 Meniscus depins as it moves over an obstacle at a substrate speed of 100 μm/s and at a blade height of 160 μm. Dry substrate (*I*) is exposed as the meniscus (*II*) moves over the obstacle (*III*). The depinning length l_{dep} is measured as the distance between the unperturbed contact line and the obstacle just after depinning (indicated in the *center image*)

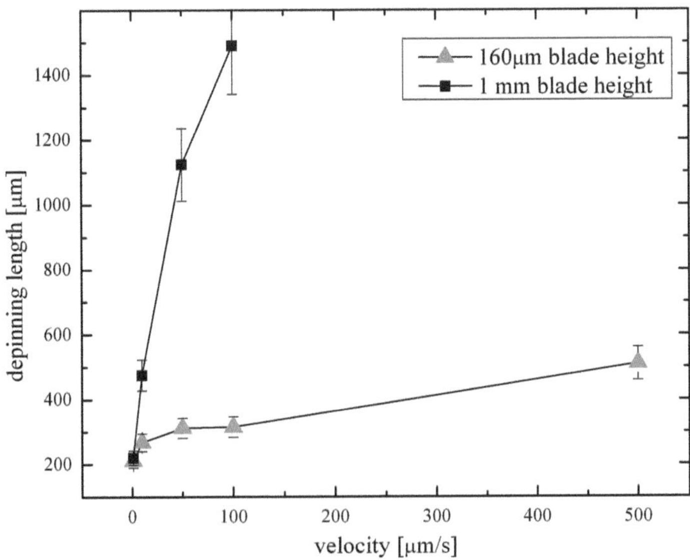

Fig. 2.11 Depinning lengths of a dynamic meniscus, measured by video microscopy as a function of withdrawal velocity for two blade heights. The lines are meant to guide the eyes

2.3.5 *Effect of Blade Height and Substrate Velocity on Deposition Homogeneity*

Theoretical considerations and the results above indicate an influence of the setup geometry on the homogeneity of deposited particle films. To characterize this influence, depositions at various blade heights and substrate velocities were made. The blade height can be set freely for any given deposition. The withdrawal velocity, however, must match the growth velocity v_f of the particle film and is tied to at least one other parameter. NAGAYAMA'S equation (Eq. 2.11) offers two ways to increase the growth velocity of a film with fixed thickness h_f: increasing the particle volume fraction in the suspension ϕ or increasing the evaporative flux j_e. The evaporative flux is adjustable via the environmental pressure [2], the environmental humidity [16] and the substrate temperature [24]. Substrate temperature is easily changed, but has unwanted side effects: Fig. 2.12 shows the increase of the contact angle with substrate temperature. Thus, increasing the evaporation flux by raising the temperature to achieve a high growth velocity and a low contact angle was not possible. Adjusting the environmental pressure and humidity is cumbersome.

Surfactants can also tune the contact angle. However, surfactants are usually non-volatile and will co-deposit on the particle film, changing its structure and chemical nature. In addition, the local variations of surfactant concentration cause MARANGONI flows which compromise the convective assembly process.

Instead, we used a high particle volume fraction ϕ of the suspension, which led to the desired increase in growth velocity v_f. Together, a moderate assembly

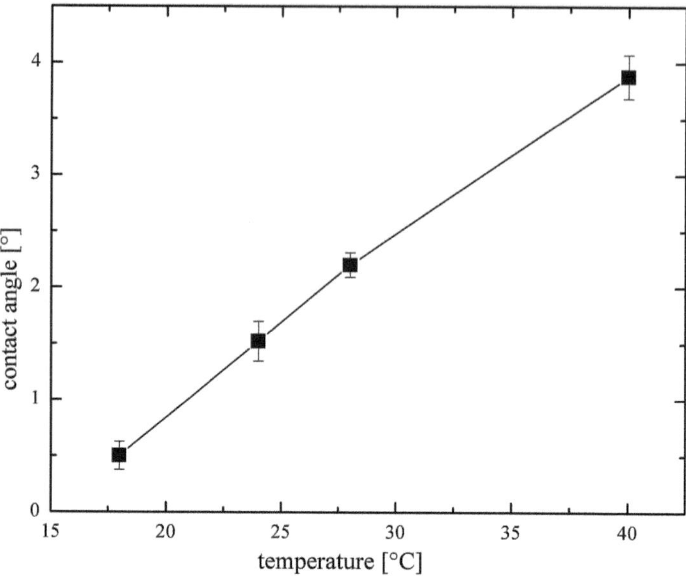

Fig. 2.12 Temperature dependence of the contact angle, measured during deposition via interference microscopy. The lines are guides to the eyes

temperature and high particle concentration prove a good strategy for particle assembly. We obtained high-quality particle monolayer from an aqueous suspension of 500 nm polystyrene particles with $c_0 = 2.6\ \%$ solid content and 1 mm blade height at room temperature using a withdrawal rate of 10 μm/s. A dynamic contact angle of $1°$ and a curvature of $8 \cdot 10^{-4}$ μm^{-1} were measured. They match the values determined for pure water menisci.

The optimized conditions were used to assess the influence of the substrate-blade distance and substrate velocity on the homogeneity of the deposited film. Thus, we reduced the blade-surface distance in one experiment and lowered the particle concentration in the other. The withdrawal velocity was chosen to produce a monolayer in all cases.

The film homogeneity was analyzed using a video scan with a low-magnification microscope objective under white light transmission illumination. Any heterogeneity in the film, such as holes, gaps or multilayer, appeared as variations in brightness. The standard deviation of the pixel gray values across the image gave an estimate for the film homogeneity. It was calculated for every frame to analyze the homogeneity over the full film length. A consistently low standard deviation over several frames indicates a homogeneous film.

Figure 2.13 shows that the film is most uniform at 1 mm blade height and high particle concentration. High initial standard deviations in the first ten millimeters are caused by static deposition before the actual process started. They are followed

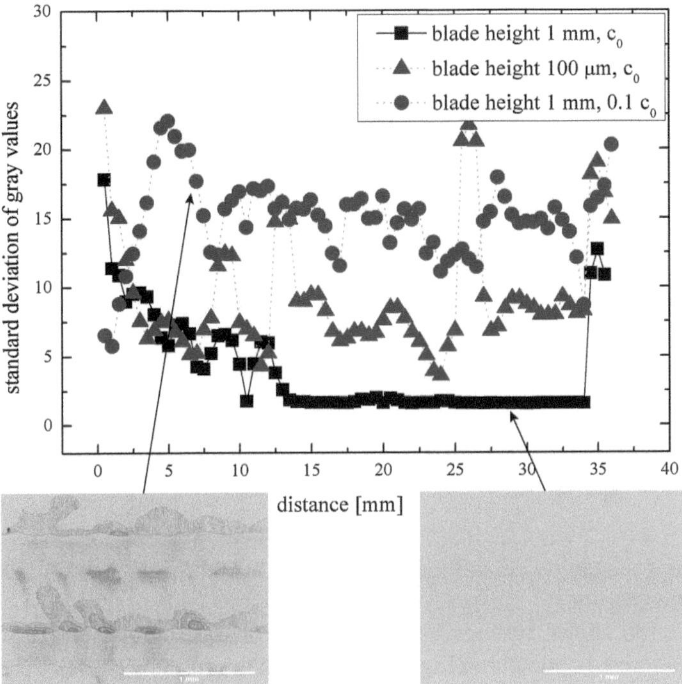

Fig. 2.13 Homogeneity analysis of films that were deposited using different conditions. A combination of large blade height and high deposition velocity (set by the high particle concentration c_0) yielded the most homogeneous film. The insets show two examples for the microscopy images that were analyzed for the graph, *scale bar* represents 1 mm

by low deviations over the rest of the 35 mm deposition length. We expect that the homogeneous film is only limited by the setup.

In contrast, deposition at a low blade-substrate distance or at low particle concentrations lead to consistently high standard deviations, indicating that these parameters were not suitable for high-quality film deposition.

2.3.6 Effect of Deposition Homogeneity on the Microscopic Quality of the Particle Film

Figure 2.14 shows a convectively assembled particle film that highlights the importance of macroscopic film homogeneity for microscopic deposition quality. The film exhibits a row of holes that formed concurrently. The polarization contrast highlights how these relatively small holes (below 100 µm in width) nucleate large regions of deviating microstructure, with the range of the perturbations extending over much longer distances than the holes themselves. Electron microscopy reveals that the

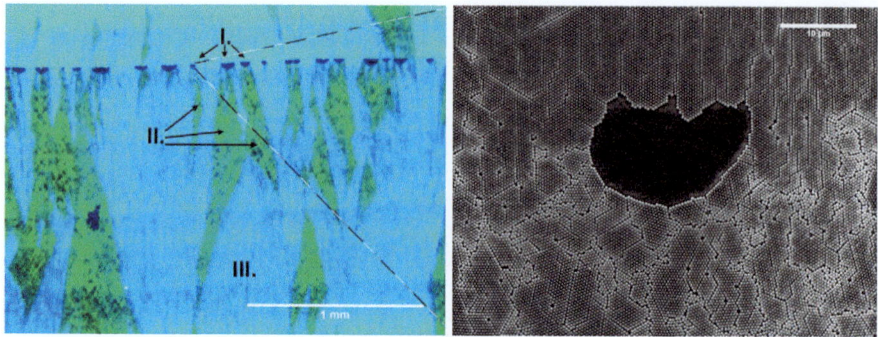

Fig. 2.14 Polarization light microscopy image and scanning electron microscopy image of a film that was convectively assembled from 500 nm diameter polystyrene particles. The dark spots (*I*) are holes in the film. They nucleate the growth of a small-grained polycrystalline film (*II*) that differs from the otherwise homogeneous single-crystal film (*III*). The film was grown from *top* to *bottom*

holes disrupt the single crystalline particle film, which becomes polycrystalline in the perturbed region.

Figure 2.15a shows two representative electron micrographs of the films that were characterized optically in Fig. 2.13. Electron microscopy proves the absence of thickness variations or holes in the film deposited under optimized conditions. The only observable defects are few single-particle voids and sub-particle diameter wide drying-cracks, not visible in optical microscopy. In the film deposited at low particle concentration the electron micrograph reveals thickness variations, small holes and grain boundaries. Figure 2.15b (I) shows a high-magnification electron micrograph of a height transition in the particle film. The nucleation of a polycrystalline region in the particle film is clearly visible behind the step edge.

The long-range translational order was investigated using FOURIER transformations of the electron micrographs. The Fourier image of the homogeneous film features reflexes that indicate long-range hexagonal order that is unaffected by the drying cracks. The Fourier image of the inhomogeneous film with thickness variations features rings that indicate a polycrystalline structure.

The local order of particle packing is characterized by the average number of nearest neighbors. For a perfect two-dimensional hexagonal c lose packing, every particle should touch 6 nearest neighbors. In a typical homogenous particle film (Fig. 2.15a) (II), we find an average number of nearest neighbors of 5.9, limited by the drying cracks. In the inhomogeneous film (Fig. 2.15a) (I) the average number of nearest neighbors is around 5.3, indicating the higher number of particles at sites with distorted local packing.

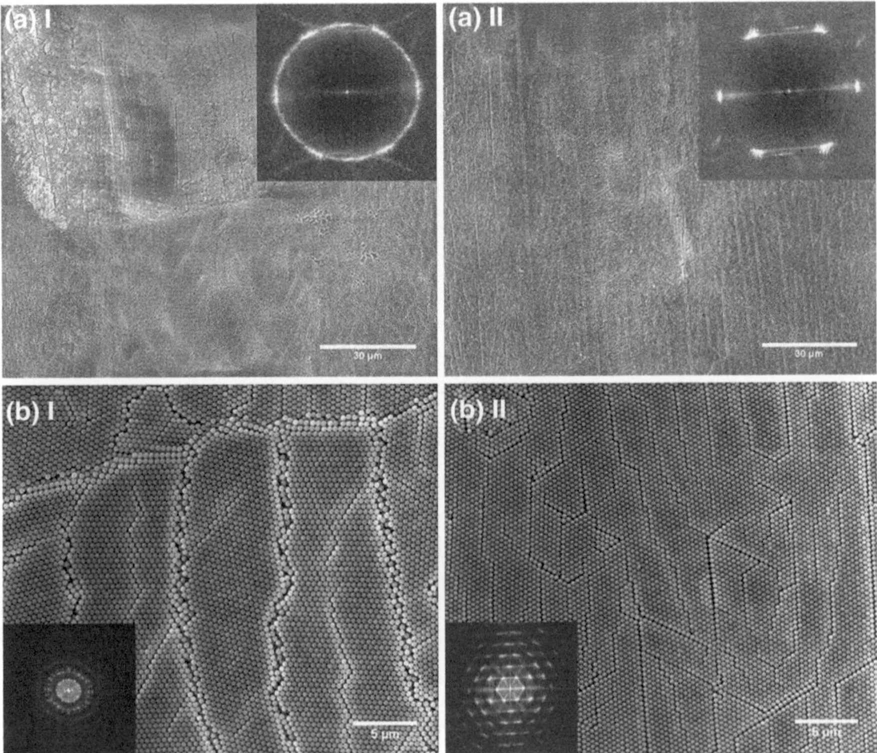

Fig. 2.15 Scanning electron micrographs of films that were convectively assembled from 500-nm-diameter polystyrene particles. The *top row* **a** shows overviews (*scale bars* correspond to 30 μm), the *bottom row* **b** shows high magnification images of the films (*scale bars* correspond to 5 μm). The films deposited from suspension with 0.1 c_0 (denoted as *I*) exhibit large inhomogeneities such as holes and double layers which cause growth of polycrystalline film. The film deposited under optimal conditions (denoted as *II*) is perfectly crystalline with microscopic voids and drying cracks. The insets show Fourier transformations that indicate long-range translational order in the film deposited under optimal conditions (*II*). All films shown here were grown from the top to the bottom of the images

2.4 Discussion and Conclusion

The meniscus shape depends on the setup and influences film deposition. We found that contact line curvature, meniscus curvature and contact angle change with blade height and substrate velocity as expected from theory. We also found that these geometrical features affect assembly:

- A contact line with small curvature provides uniform evaporation over the substrate width.
- A flat meniscus has a weak tendency to jump between particle film heights during deposition.

- Contact angle and meniscus curvature together govern the deformability of the meniscus. A deformable, slack meniscus can accommodate different deposition conditions, prevents defects from spreading and has a low tendency to de-pin and form holes in the particle film. The deposition thus becomes robust against the ubiquitous perturbations of process conditions.

We used optimized conditions to deposit homogeneous films over five square centimeters that were limited only by the width of the substrate and the travel distance of the stage. The quality of these films was compared to other films in terms of optical homogeneity, long-range and short-range order using a combination of optical and electron microscopy. Under optimized conditions the films grew out initial defects and turned into single crystalline particle monolayer. Under other conditions the density of defects stayed high and the crystalline structure was continuously disrupted.

All deposition setups face a trade-off between robustness and homogeneity of the deposition. A blade is advantageous because it straightens the contact line, but it imposes a curvature on the meniscus and on the contact line at the sides of the substrate. Setups without blade have menisci with minimal curvatures. However, in such setups the curvature of the contact line will extend over the entire substrate and the evaporation rate will vary over the substrate. We expect an optimum for setups where the blade-substrate distance is close to the capillary length such that meniscus curvature is small, but the contact line is straight.

A moving substrate aids deposition in all setups because it provides a low, dynamic contact angle and the film thickness can be controlled by the withdrawal velocity. We suggest studying high withdrawal velocities in setups with a confined meniscus. Balance then requires an increased influx of particles, for example due to increased particle concentration or evaporation rate. The low dynamic contact angle and the flat meniscus at high velocities will contribute to an increased homogeneity and stability of the deposition. At very high withdrawal velocities (say, above 100 μm/s) viscous dissipation deforms the meniscus, which becomes long and flat as we have shown. This geometry should be optimal for convective particle assembly according to above discussion. However, effects not considered in this study such as hydrodynamic instabilities occurring at that high velocity might be limiting [42].

2.5 Experimental

All experiments were performed in a climate-controlled laboratory at a temperature of 22 ± 1°C and a relative humidity of 40 ± 5%.

2.5.1 Substrates and Particles

Substrates for the measurement of meniscus profiles and for the deposition of particle films were either silicon wafers (Asahi Kasei, Tokyo, Japan, $<100>$-oriented) with a native oxide layer or standard microscopy slides (Marienfeld, Lauda-Königshofen, Germany, pure white glass), if transparent substrates were required. The substrates were cleaned in an ultrasonic bath using isopropanol (Sigma-Aldrich, Deisenhofen, Germany, puriss. p. a.) and deionized water (ultrapure quality from a Millipore unit). Immediately before use, the substrates were hydrophilized by oxygen plasma treatment at a pressure of 0.3 mbar and RF power of 100 W for 5 minutes in the quartz tube of a commercial plasma reactor (low-pressure reactor PICO, RF source at 13.56 MHz, Diener electronic, Ebhausen, Germany). All substrates were spontaneously wetted by water after this step.

Aqueous suspensions of emulsion-polymerized polystyrene microspheres with specified diameters of 457 ± 10 nm were obtained from Polysciences (Polysciences Europe GmbH, Eppelheim, Germany), with a solid content of 2.6 %. The concentration was adjusted by dilution with ultrapure water. Before deposition, the suspensions were filtered by pressing them through 0.8 μm cellulose acetate syringe filters (Whatman, Dassel, Germany).

2.5.2 Setup

The setup used in this study is shown schematically in Fig. 2.16. It provides a restricted meniscus [28]. We used this setup geometry because its meniscus has well-defined geometry and is accessible to interference microscopy. The geometrical constraints (blade height) and the substrate velocity were varied in wide intervals, allowing extrapolations of the results to setups with static substrates or without geometrical constraints. All substrates were fixed on a temperature-controlled sample holder that could be moved by a motorized translational stage. Temperature was controlled within $\pm 0.5\,^{\circ}$C by a refrigerated circulator (Lauda RE 204, Lauda-Königshofen, Germany) and measured using a thermocouple mounted on the holder. The substrate was left to equilibrate on the holder for at least 5 minutes after each temperature change. For experiments with evaporation barriers, the substrate was enclosed in a Teflon trough with side walls of 2 mm height. The blade holding the liquid reservoir was mounted such that its distances to the substrate could be varied. A volume of roughly 100 μl of liquid was placed between the blade and the substrate. Relative motion between the blade and the substrate was controlled by a precision linear stage (PLS-85, Micos, Eschbach, Germany), which assured steady motion of the stage under the blade (a maximal deviation of 30 nm/s from the target velocity was measured by the manufacturer via interferometry).

The entire setup was mounted on the stage of a conventional optical microscope (AxioImager.A1m, Zeiss, Oberkochen, Germany), which enabled video microscopy

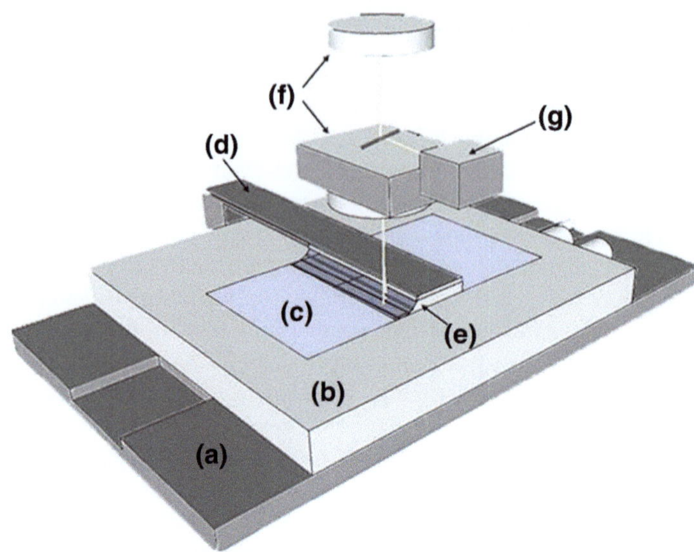

Fig. 2.16 Experimental setup. A linear stage with stepper motor **a** moves a temperature-controlled sample holder **b** on which the substrate **c** is clamped under a blade **d** which holds the suspension reservoir **e**. The process is observed using a light microscope with CCD camera **f** under the illumination of a blue LED **g**

and interference microscopy, but limited the travel distance of the linear stage to 35 mm. Videos were taken using a greyscale high-speed camera (Zeiss AxioCam) and evaluated using the Zeiss AxioVision software. Interference microscopy was performed using the same setup with a blue high-power LED (460 nm emission, FWHM 20 nm) for illumination.

2.5.3 Interference Microscopy

We used optical microscopy in reflection mode under the illumination of monochromatic light to obtain interference microscopy images. The light reflected from the substrate and the gas-liquid interface created interference patterns that were depending on the local thickness of the liquid, which were recorded using the microscope's camera. Line scans in x-direction over the images displayed characteristic brightness oscillations with minima and maxima related to the thickness h_m of the liquid as in,

$$h_m = \frac{m\lambda}{4n_{water}},$$ (2.14)

where λ is the wavelength of the light, $n_{water} = 1.33$ the refractive index of water and m the order of the interference, where odd m give dark fringes and even m give bright fringes. The brightness oscillations in the microscope image were analyzed using this equation to reconstruct the shape of the gas-liquid interface in the $x - z$-plane to an accuracy of 4 %. The error originates from uncertainties in defining the exact position of the fringes in the interference pattern during analysis.

After reconstructing the shape of the gas-liquid interface of a meniscus, we fitted the obtained profile to a second order polynomial $f(x)$ with a regression coefficient better than 0.999. The first interference minimum for pure solvent is expected at a film thickness of 86.47 nm. During deposition, the particles' diameter sets the minimum thickness. Thus, the first observable interference minimum during deposition was at 605.26 nm. The contact angle Θ of the meniscus on the substrate could then be derived from the slope of the profile at the intersection with the x-axis using following equation:

$$\Theta = \arctan\left[\left. \frac{\mathrm{d}f(x)}{\mathrm{d}x} \right|_{h_m=0} \right].$$
(2.15)

The height profile also yields the curvature of the gas-liquid interface, which we calculated using

$$\frac{1}{R} = \frac{\frac{\mathrm{d}^2 f(x)}{\mathrm{d}x^2}}{\left[1 + \left(\frac{\mathrm{d}f(x)}{\mathrm{d}x}\right)^2\right]^{\frac{3}{2}}}.$$
(2.16)

Error bars indicate the scattering of the experimental values for contact angles and curvatures measured in the same region of the meniscus around the given average.

2.5.4 Video Microscopy

The depinning length was measured using a small droplet of the epoxy glue that was placed on the substrate using a needle tip, followed by the standard procedure for substrate hydrophilization. The substrate was fixed on the sample holder and was withdrawn from a water reservoir with different withdrawal velocities and at different blade heights. After calibration, the depinning distances were measured in the Zeiss AxioVision software.

For film quality analysis the video scans of the particle films were saved from the AxioVision software as image sequences and loaded with ImageJ [43]. After background subtraction (rolling ball algorithm, light background, 50 pixel radius, [44]) the standard deviation of the grey values of the pixels for each frame was calculated.

Videos were taken with a 5x magnification objective which corresponds to frame sizes of 2.08×1.56 mm^2.

2.5.5 *Polarization Light Microscopy and Scanning Electron Microscopy*

Polarization light microscopy of particle films on standard microscope slides was performed using a Leitz Orthoplan-Pol microscope (Ernst Leitz Wetzlar GmbH, Wetzlar, Germany) with transmission illumination. Electron micrographs were obtained using a FEI Quanta 400F scanning electron microscope (FEI Europe, Eindhoven, Netherlands). A thin gold film was sputtered on the particle films to minimize charging effects during electron microscopy. Electron micrographs were transformed by the Fast Fourier Transformation algorithm provided by the ImageJ software. Additionally, the number of nearest neighbors was calculated using the particle tracking algorithm published in [45] to extract the individual particles' coordinates. Roughly 6500 particles per image were then analyzed using a simple script which counts all particles within a distance of two radii as nearest neighbors. The average number of nearest neighbors was then calculated by averaging the results.

References

1. N.D. Denkov, O.D. Velev, P.A. Kralchevsky, I.B. Ivanov, H. Yoshimura, K. Nagayama, Mechnism of formation of two-dimensional crystals from latex particles on substrates. Langmuir **8**, 3183–3190 (1992)
2. T.P. Rivera, O. Lecarme, J. Hartmann, E. Rossitto, K. Berton, D. Peyrade, Assisted convective-capillary force assembly of gold colloids in a microfluidic cell: plasmonic properties of deterministic nanostructures. J. Vac. Sci. Technol. B: Microelectron. Nanometer Struct. **26**(6), 2513–2519 (2008)
3. X. Checoury, S. Enoch, C. Lopez, A. Blanco, Stacking patterns in self-assembly opal photonic crystals. Appl. Phys. Lett. **90**, 161131 (2007)
4. P. Jiang, J.F. Bertone, K.S. Hwang, V.L. Colvin, Single-crystal colloidal multilayers of controlled thickness. Chem. Mater. **11**, 2132–2140 (1999)
5. E. Vekris, V. Kitaev, D.D. Perovic, J.S. Aitchinson, G.A. Ozin, Visualization of stacking faults and their formation in colloidal photonic crystal films. Adv. Mater. **20**, 1110–1116 (2008)
6. S.G. Romanov, U. Peschel, M. Bardosova, S. Essig, K. Busch, Suppression of the critical angle of diffraction in thin-film colloidal photonic crystals. Phys. Rev. B: Condens. Matter Mater. Phys. **82**, 115403 (2010)
7. P. Kumnorkaew, Y.-K. Ee, N. Tansu, J.F. Gilchrist, Investigation of the deposition of microsphere monolayers for fabrication of microlens arrays. Langmuir **24**, 12150–12157 (2008)
8. M. Bunzendahl, P.L.-V. Schaick, J.F.T. Conroy, C.E. Daitch, P.M. Norris, Convective self-assembly of stoeber sphere arrays for syntactic interlayer dielectrics. Colloids Surf. A: Physicochem. Eng. Aspects **182**, 275–283 (2001)
9. B.G. Prevo, E.W. Hon, O.D. Velev, Assembly and characterization of colloid-based antireflective coatings on multicrystalline silicon solar cells. J. Mater. Chem. **17**, 791–799 (2007)
10. K.W. Kho, Z.X. Shen, H.C. Zeng, K.C. Soo, M. Olivo, Deposition method for preparing SERS-active gold nanoparticle substrates. Anal. Chem. **77**(22), 7462–7471 (2005)
11. B.G. Prevo, D.M. Kuncicky, O.D. Velev, Engineered deposition of coatings from nano-andmicro-particles: a brief review of convective assembly at high volume fraction. Colloids Surf. A: Physicochem. Eng. Aspects **311**, 2–10 (2007)

12. N.H. Finkel, B.G. Prevo, O.D. Velev, L. He, Ordered silicon nanocavity arrays in surface-assisted desorption/ionization mass spectrometry. Anal. Chem. **77**, 1088–1095 (2005)
13. A.D. Ormonde, E.C.M. Hicks, J. Castillo, R.P.V. Duyne, Nanosphere lithography: Fabrication of large-area Ag nanoparticle arrays by convective self-assembly and their characterization by scanning UV-Visible extinction spectroscopy. Langmuir **20**, 6927–6931 (2004)
14. G.S. Lozano, L.A. Dorado, R.A. Depine, H. Miguez, Towards a full understanding of the growth dynamics and optical response of self-assembled photonic colloidal crystal films. J. Mater. Chem. **19**, 185–190 (2009)
15. C.D. Dushkin, G.S. Lazarov, S.N. Kotsev, H. Yoshimura, K. Nagayama, Effect of growth conditions on the structure of two-dimensional latex crystals: experiment. Colloid Polym. Sci. **277**, 914–930 (1999)
16. B.G. Prevo, O.D. Velev, Controlled, rapid deposition of structured coatings from micro- and nanoparticle suspensions. Langmuir **20**, 2099–2107 (2004)
17. H.J. Schöpe, A.B. Fontecha, H. König, J.M. Hueso, R. Biehl, Fast microscopic method for large scale determination of structure, morphology, and quality of thin colloidal crystals. Langmuir **22**, 1828–1838 (2006)
18. L. Meng, H. Wei, A. Nagel, B.J. Wiley, L.E. Scriven, D.J. Norris, The role of thickness transitions in convective assembly. Nano Lett. **6**(10), 2249–2253 (2006)
19. J. Hilhorst, V.V. Abramova, A. Sinitskii, N.A. Sapoletova, K.S. Napolskii, A.A. Eliseev, D.V. Byelov, N.A. Grigoryeva, A.V. Vasilieva, W.G. Bouwman, K. Kvashnina, A. Snigirev, S.V. Grigoriev, A.V. Petukhov, Double stacking faults in convectively assembled crystals of colloidal spheres. Langmuir **25**, 10408–10412 (2009)
20. H. Cao, D. Lan, Y. Wang, A.A. Volinsky, L. Duan, H. Jiang, Fracture of colloidal single-crystal films fabricated by controlled vertical drying deposition. Phys. Rev. E: Stat. Nonlin. Soft Matter Phys. **82**, 031602 (2010)
21. D.J. Norris, E.G. Arlinghaus, L. Meng, R. Heiny, L.E. Scriven, Opaline photonic crystals: how does self-assembly work? Adv. Mater. **16**, 1393–1399 (2004)
22. D.D. Brewer, J. Allen, M.R. Miller, J.M. de Santos, S. Kumar, D.J. Norris, M. Tsapatsis, L.E. Scriven, Mechanistic principles of colloidal crystal growth by evaporation-induced convective steering. Langmuir **24**, 13683–13693 (2008)
23. D. Gasperino, I. Meng, D.J. Norris, J.J. Derby, The role of fluid flow and convective steering during the assembly of colloidal crystals. J. Cryst. Growth **310**, 131–139 (2008)
24. L. Malaquin, T. Kraus, H. Schmid, E. Delamarche, H. Wolf, Controlled particle placement through convective and capillary assembly. Langmuir **23**, 11513–11521 (2007)
25. R.D. Deegan, O. Bakajin, T.F. Dupont, G. Huber, S.R. Nagel, T.A. Witten, Capillary flows as the cause of ring stains from dried liquid drops. Nature **389**, 827–829 (1997)
26. A.D. Dimitrov, A. Nagayama, Continuous convective assembling of fine particles into two-dimensional arrays on solid surfaces. Langmuir **12**, 1303–1311 (1996)
27. S. Watanabe, K. Inukai, S. Mizuta, M.T. Miyahara, Mechanism for stripe pattern formation on hydrophilic surfaces by using convective self-assembly. Langmuir **25**, 7287–7295 (2009)
28. K. Chen, S.V. Stoianov, J. Bangerter, H.D. Robinson, Restricted meniscus convective self-assembly. J. Colloid Interface Sci. **344**, 315–320 (2010)
29. J. Kleinert, S. Kim, O.D. Velev, Electric-field-assisted convective assembly of colloidal crystal coatings. Langmuir **26**, 10380–10385 (2010)
30. M. Sujanani, P.C. Wayner, Transport processes and interfacial phenomena in an evaporating meniscus. Chem. Eng. Commun. **118**, 89–110 (1992)
31. T. Young, An Essay on the Cohesion of Fluids. Philosophical Transactions of the Royal Society of London, **95**, 65–87 (1805)
32. R. Eötvös, Ueber den Zusammenhang der Oberflächenspannung der Flüssigkeiten mit ihrem Molecularvolumen. Ann. Phys. **263**(3), 448–459 (1886)
33. P.G. de Gennes, F. Brochard-Wyart, D. Quéré, *Capillarity and Wetting Phenomena* (Springer Science+Bussiness Media, LLC, 2004)
34. P.D. de Gennes, Wetting: statics and dynamics. Rev. Mod. Phys. **57**(3), 827–863 (1985)

35. P.G. de Gennes, X. Hua, P. Levinson, Dynamics of wetting: local contact angles. J. Fluid Mech. **212**, 55–63 (1990)
36. J.F. Joanny, P.G. de Gennes, A model for contact angle hysteresis. J. Chem. Phys. **81**(1), 552–562 (1984)
37. E. Raphael, P.G. de Gennes, Dynamics of wetting with nonideal surfaces. The single defect problem. J. Chem. Phys. **90**(12), 7577–7584 (1989)
38. R.D. Deegan, O. Bakajin, T.F. Dupont, G. Huber, S.R. Nagel, T. Witten, Contact line deposits in an evaporating drop. Phys. Rev. E: Stat. Nonlin. Soft Matter Phys. **62**(1), 756–765 (2000)
39. R.D. Deegan, Pattern formation in drying drops. Phys. Rev. E: Stat. Nonlin. Soft Matter Phys. **61**(1), 475–485 (2000)
40. E.C.H. Ng, K.M. Chin, C.C. Wong, Controlling inplane orientation of a monolayer colloidal crystal by meniscus pinning. Langmuir **27**, 2244–2249 (2011)
41. G.D. Nadkarni, S. Garoff, An investigation of microscopic aspects of contact angle hysteresis: pinning of the contact line on a single defect. Europhys. Lett. **20**(6), 523–528 (1992)
42. E. Adachi, A.S. Dimitrov, K. Nagayama, Stripe patterns formed on a glass surface during droplet evaporation. Langmuir **11**, 1057–1060 (1995)
43. ImageJ.http://rsb.info.nih.gov/ij
44. S.R. Sternberg, Biomedical image processing. Computer **16**, 22–34 (1983)
45. I.F. Sbalzarini, P. Koumoutsakos, Feature point tracking and trajectory analysis for video imaging in cell biology. J. Struct. Biol. **151**, 182–195 (2005)

Chapter 3
Crystallization Mechanisms in Convective Particle Assembly

Abstract Colloidal particles are continuously assembled into crystalline particle coatings using convective fluid flows. Assembly takes place inside a meniscus on a wetting reservoir. The shape of the meniscus defines the profile of the convective flow and the motion of the particles. We use optical interference microscopy, particle image velocimetry and particle tracking to analyze the particles' trajectory from the liquid reservoir to the film growth front and inside the deposited film as a function of temperature. Our results indicate a transition from assembly at a static film growth front at high deposition temperatures to assembly in a precursor film with high particle mobility at low deposition temperatures. A simple model that compares the convective drag on the particles to the thermal agitation explains this behavior. Convective assembly mechanisms exhibit a pronounced temperature dependency and require a temperature that provides sufficient evaporation. Capillary mechanisms are nearly temperature independent and govern assembly at lower temperatures. The model fits the experimental data with temperature and particle size as variable parameters and allows prediction of the transition temperatures. While the two mechanisms are markedly different, dried particle films from both assembly regimes exhibit hexagonal particle packings. We show that films assembled by convective mechanisms exhibit greater regularity than those assembled by capillary mechanisms.

3.1 Introduction

Two-dimensional lattices of sub-micron particles are useful surface coatings: they induce well-defined roughnesses, curvatures and periodicities to the interface. Surface roughness affects the interaction with biological systems like cells [1, 2] and arthropods [3] and the contact angles on superhydrophobic [4] and self-cleaning surfaces [5]. Curvature is an important parameter for magnetic high density data storage [6] or microlens arrays [7, 8]. Periodic surfaces modulate sound and light e.g. in planar waveguides [9], diffraction gratings [10], coatings with bandgaps for

P. G. Born, *Crystallization of Nanoscaled Colloids*, Springer Theses,
DOI: 10.1007/978-3-319-00230-9_3, © Springer International Publishing Switzerland 2013

certain frequencies [11] and in anti-reflective coatings [12]. Metal particle layers provide plasmonic wave guides [13]. Particle films can be used as sacrificial layer in particle lithography [14, 15]. Most of the applications exploit the tendency of the particles to form regular close-packed structures, often referred to as colloidal crystallization [16].

A convenient bottom-up fabrication process of colloidal crystals is the convective particle assembly technique [10]. Large-area films with a high degree of order can be deposited by convective assembly in simple experimental setups [17–24]. The basic principle of convective assembly is the transport of particles from a reservoir into a meniscus that forms on a wetting substrate. Particles are confined at the contact line and deposited onto the substrate. When the substrate is slowly withdrawn, the meniscus moves and the accumulated particles are continuously deposited as a film. Crystallization occurs in a transition region between the meniscus and the dried particle film (see Fig. 3.1).

Macroscopically, convective assembly relies on a solvent flow that increases the particle concentration in the meniscus at the rim of the reservoir [10]. In the beginning, evaporation is enhanced by the divergence of the evaporation at the contact line of the meniscus [25, 26]. Particles are driven towards the three-phase boundary line, where they accumulate. The accumulated particles eventually pin the boundary line. If the meniscus moves—either due to the reduction of the liquid volume or due to active motion of the substrate relative to the liquid reservoir—a liquid film is drawn out. Evaporation continuously thins this film. When its thickness drops below the particle diameter, the liquids surface is deformed by the particles, which increases the pressure gradient from the meniscus into the wet particle film [10]. The particles are then pulled together by capillary forces. If the velocity of the meniscus is equal to the growth of the particle film, continuous deposition ensues—convective particle assembly sets in. The deposited film stays wet; its increasing surface area increases the evaporation rate, the convective flow rate, and the particle deposition rate. At a certain distance l_e from the three-phase boundary line, the particle film dries in completely. When the deposited film becomes longer than l_e, convective assembly enters a steady state where evaporation rate, convective flow rate and particle deposition rate stay constant. Only macroscopic defects or thickness variations in the film disturb the flow profile and lead to a breakdown of the steady-state particle deposition (see Chap. 2).

Microscopically, regular particle assembly requires guiding the particles to lattice positions. Transport towards the film is necessary, but insufficient, to create order. Two well-developed theories describe the microscopic assembly mechanisms that arrange particles during convective particle assembly: convective steering [27–29] and capillary crystallization [16, 30, 31].

The *capillary crystallization* model was developed by NAGAYAMA et al. to explain the assembly of particles in thinning liquid films [16] (see Fig. 3.1c). The authors identified an attractive capillary interaction, the "immersion force", between particles that are deforming the liquid surface. It is much stronger than thermal agitation even for very small particles down to a few nanometers if a substrate supports the particles during deformation of the liquid surface [32]. Under this strong attraction, small hexagonal packed nuclei form and grow into continuous polycrystalline,

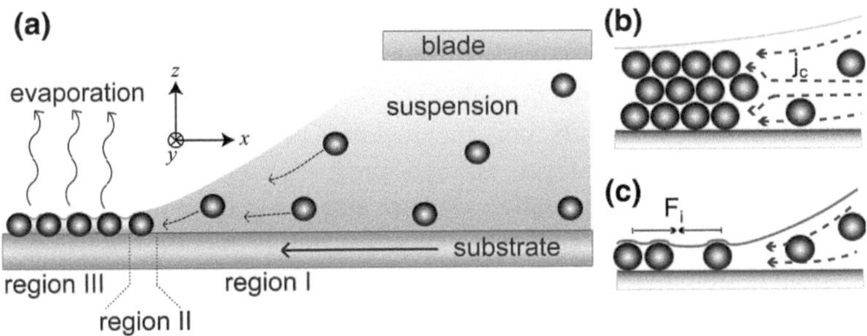

Fig. 3.1 **a** The three regions of convective particle assembly: In region I, the meniscus, free particles are convectively transported to the particle film. In region II, the particles are arranged to the particle film structure. In region III, the particles form a wet particle film with pronounced evaporation. A deposition in blade coating geometry as used in the experiments is shown. In this geometry a blade fixes the reservoir and straightens the meniscus. **b** and **c** illustrate the crystallization mechanisms. **b** Scheme of convective steering. The particles are transported by the convective solvent flow J_c into niches that define a crystalline packing. **c** Scheme of capillary crystallization. The particles are transported into a thin solvent film. The deformation of the liquid surface induces attractive immersion forces F_i, which drag the particles into a crystalline packing

two-dimensional, close-packed particle films. Convective flow is neglected in this model of the assembly process, it merely ensures transport of particles from a reservoir to the assembly region of the liquid film. Capillary crystallization processes are temperature-dependent only in so far as the solvent's surface tension depends on temperature. The surface tension of water drops almost linearly in the relevant temperature range, by $\approx 6\%$ between 10 to 40 °C [33].

The *convective steering* mechanism depends on the exact motion of the solvent (see Fig. 3.1b). NORRIS et al. performed a network analysis of the fluid flow through a regular arrangement of particles to explain the striking yield of face-centered cubic particle packings that occur in the convective assembly of bulk crystals and far surpass the fraction of hexagonal close-packed or random close-packed particle arrangements [29]. They assumed that previously deposited particles form a scaffold for the placement of incoming particles. The solvent flows into the niches of this scaffold and drags the particles with them. Particles are predominantly transported to and fixed in niches with the strongest influx. The authors showed that niches without underlying particles have the greatest influx, which favors particle stacking in an ABCABC...-sequence, the structure of face-centered cubic crystals. The temperature dependence of this mechanism stems from the evaporation process. The evaporation rate is proportional to the vapor pressure of the solvent which exponentially depends on temperature. The vapor pressure of water drops by $\approx 84\%$ between 40 and 10 °C [33].

It is yet unknown which of these mechanisms prevails in the convective assembly of two-dimensional crystalline films from a meniscus. Interference microscopy has shown that the liquid film at the film growth front is slightly thicker than the particle diameter Chap. 2, which indicates assembly without deformation of the liquid surface and suggests convective steering. As the film dries, however, the particle arrangement often changes, a process surely related to capillary crystallization. Our goal in this study is to clarify the role of both mechanisms in the deposition of regular monolayer.

A better understanding of the assembly mechanism will aid the control of convective particle assembly. Presently, only the thickness of the particle film [17, 23] and the areal fraction of sub-monolayer [8] can be precisely controlled in pure convective assembly. Control over the crystalline domain size is a subject of ongoing research [24, 34, 35]. Moreover, little control of layer structure is possible today. One would like to switch between amorphous and crystalline packings or between fine-grained and coarse-grained crystalline packings during assembly because amorphous films can tailor optical properties [36] and have found applications as anti-reflective coatings [5] and in Bragg mirrors [37]. Adjustment of the dielectric constant of particle films [38] or the structures formed by colloidal lithography [14] is also desirable. Sufficient structural control over the films presently requires the use of different particles (e.g. low and high size dispersities [17]) or templates [20].

In the next section, we develop a simple model to estimate the work required to move particles from their position back into the bulk suspension. We divide the convective particle assembly process into three regions (see Fig. 3.1): Region III is the capillary pump, a wet particle film. Region II is the rim of the liquid meniscus, the region where particles are confined and are attached to the film and where the structure formation takes place. Region I is the liquid meniscus in which the particles are transported from the reservoir to the rim. The action of the capillary pump has already been analyzed before [23]. The shape of the meniscus in region I and its effects on film formation is also known [8] (compare Chap. 2). In this paper, we focus on the processes in region II. Our model describes how much work is required to remove particles from this region to the bulk suspension.

Random thermal agitation counteracts the directing forces that cause convective particle assembly. Successful assembly requires sufficient transport to the particle film and a force that is strong enough to hold the particles in their lattice positions until they loose mobility. Thermal agitation is linearly temperature-dependent. The different temperature dependencies of the two assembly mechanisms described above allow to distinguish their roles. The exponential decay of the evaporation rate and the linear behavior of the thermal energy predict a regime where convective steering fails while capillary interactions still lead to regular assembly. In the next section, we compare the thermal energy scale to the work required to move a particle from its position in the convective assembly region. Then we experimentally determine the liquid flow profile at different temperatures, analyze the temperature dependency and identify the assembly mechanism.

3.2 Model

Thermal agitation works against the forces exerted by the convective flow. Here we develop a model for the flow profile, calculate the drag force on the particles and the work required to move a particle quasi-statically against the flow. Other interactions are neglected. In steady state, the force field generated by the fluid flow becomes conservative, and an equivalent potential field is introduced. This representation is particularly convenient when analyzing the strength of the particles' attachment to their lattice position in the growing film in Sect. 3.2.2.

3.2.1 Transport

We use Cartesian coordinates with the x-axis parallel to the substrate pointing towards the reservoir, the y-axis parallel to the growth front and the z-axis perpendicular to the substrate (Fig. 3.1). The deposition process is approximated by a quasi-static two-dimensional problem in the x-z-plane. The particle film is withdrawn from the meniscus at the same rate as the film grows so that the relative position of the growth front and l_e are constant. Note that the finite curvature of the three-phase contact line in the actual assembly leads to slightly varying geometries of the x-y-plane; they are neglected here.

Figure 3.2 illustrates the modeling steps that lead from the meniscus geometry to the velocity profile, the drag force, and, finally, the work required to drag a particle back through the convective stream into the reservoir. We start with the parabolic height profile $h(x)$ of the meniscus' liquid-air interface above the substrate at a distance x from the growth front (see Chap. 2):

$$h(x) = a' + b' \cdot x + c' \cdot x^2. \tag{3.1}$$

Volume conservation implies an inverse parabolic shape of the z-averaged velocity of the liquid in the meniscus:

$$v(x, T) = -\frac{1}{a(T) + b \cdot x + c \cdot x^2}. \tag{3.2}$$

The actual flow profile is parabolic in z-direction due to the non-slip boundary condition imposed by the substrate. However, the particle size in most of the experiments is comparable to the height of the meniscus at least up to a few particle diameter distance to the film growth front. The particles thus only experience an effective flow, in which the gradient in z-direction is averaged out. We thus omit the z-component of the flow profile. Note that our choice of coordinates leads to negative velocities of liquid flowing into the film. Parameter b and c depend on the setup geometry, in particular on the position of the blade's edge that pins the meniscus (Chap. 2). Parameter a is equal to the inverse fluid velocity at the film growth front at $x = 0$. The fluid's flow rate is proportional to the evaporation rate from the particle film [10], which we assume to be proportional to the vapor pressure of the solvent here. The vapor pressure follows the law of CLAUSIUS- CLAPEYRON, which leads to an exponential dependency of the solvent velocity on the temperature for a fixed liquid geometry:

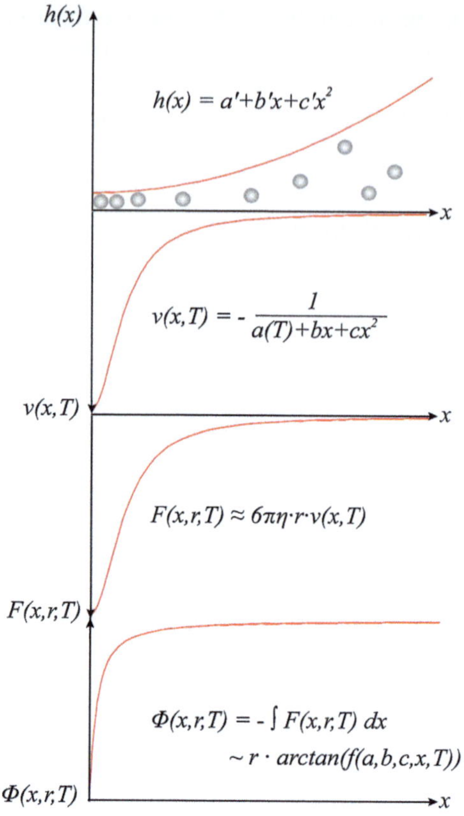

Fig. 3.2 Modeling steps from the meniscus geometry to the work required to move a particle back into the reservoir. The *parabolic shape* of the meniscus $h(x)$ impresses an inverse parabolic velocity profile $v(x, T)$. Using STOKES drag the force profile $F(x, r, T)$ is derived. Finally, by integration over the force the work $\Phi(x, r, T)$ is calculated. See text for details

$$v(0, T) = -\frac{1}{a(T)} \approx d \cdot e^{-\frac{e}{T}}. \tag{3.3}$$

The particles transported by the convective flow have negligible inertia (REYNOLDS numbers and STOKES numbers are both on the order of 10^{-4} for the measured velocities). We can therefore use their velocities (obtained from particle image velocimetry, Sect. 3.3) to estimate the solvent velocity. A fit of the particle velocities to Eq. 3.2 yields the parameters a, b and c. Measurements a different temperatures provide $a(T)$, which can be fit to Eq. 3.3 and yields the parameters d and e.

A particle with radius r that is held at a constant position experiences STOKES drag exerted by the moving solvent:

$$F(v, x, r, T) = 6\pi \eta \cdot r \cdot v(x, T), \tag{3.4}$$

where η is the solvent's viscosity. The work required to move a particle quasi-static from the film edge back into the reservoir therefore equals

$$\Phi(x, r, T) = -\int_0^\infty F(x, r, T)dx = 6\pi\eta \cdot r \cdot \int_0^\infty \frac{1}{a(T) + b \cdot x + c \cdot x^2}dx. \quad (3.5)$$

The integral in Eq. 3.5 over the inverse parabolic function can be solved analytically [39] to yield

$$\int \frac{1}{a(T) + b \cdot x + c \cdot x^2}dx = \frac{2}{\sqrt{4\,ca(T) - b^2}} \arctan\left(\frac{2\,cx + b}{\sqrt{4\,ca(T) - b^2}}\right). \quad (3.6)$$

The asymptotic behavior of the arc tangent, $\arctan(x) \xrightarrow{x \to \infty} \frac{\pi}{2}$, and $\arctan(0) = 0$ ensure a vanishing force far away from the assembly region in the reservoir and the normalization of the potential.

The solvent flow causes the transport of particles from the reservoir to the assembly region. This convective transport surpasses diffusive transport if the work required to move particles away from the assembly region is larger than their thermal agitation:

$$\Phi_t(x, r, T) = -\int_0^\infty F(x, r, T)dx > \frac{3}{2}\,kT, \quad (3.7)$$

where k is BOLTZMANN's constant and T the absolute temperature. Within the assumptions made in the derivation, the force $F(x)$ created by the fluid flow being one-dimensional and conservative, the work of removal $\Phi_t(x, r, T)$ defines a potential. This implies a considerable abstraction of the assembly process. The effects of the convective fluid flow are split into a potential that forces the particles into the particle film, and a viscous medium that dissipates the particles kinetic energy and acts as a heat bath. This abstraction makes the similarities between convective particle assembly and particle assembly in external potentials clearer. Assembly of particles by sedimentation, for example, requires the external gravitation to transport and confine the particles, and the solvent to provide dissipation and thermal agitation of the particles. We will use the terms "potential energy" or "potential" in the following to acknowledge the mathematical analogy and to emphasize the similarity between convective particle assembly and particle assembly by sedimentation or other external potentials [40].

3.2.2 Assembly

If the potential difference between reservoir and assembly region is smaller than $\frac{3}{2}\,kT$, particles can diffuse back into the reservoir. However, even a much larger potential difference does not guarantee regular deposition of the particles. Microscopic ordering requires microscopic potential energy minima, as we show in the following.

The force generated by the flow field in the above proposed model only acts in x direction and does not affect the lateral displacement of the particles. Without any further constraints, the particles would be deposited in a random manner. In case of capillary crystallization, the attractions by neighboring particles form local potential energy minima that bind and arrange the particles [16]. When convective steering governs the assembly, the required constraints originate from particles already attached to the growth front. They form 'niches' into which the free particles are transported by the solvent flow [27]. We can extend our model to account for this mechanism.

The energy required to remove a particle from its niche must be larger than thermal agitation to ensure regular arrangement. If it was smaller, the film growth front would be constantly reconfigured (see Fig. 3.3), and the convective steering must break down. We define the 'binding potential' as the work needed to move a particle one diameter away from the film growth front:

$$\Phi_b(x, r, T) \equiv -\int_0^{2r} F(x, r, T)dx. \tag{3.8}$$

Regular assembly can only take place if the binding potential $\Phi_b(x, r, T)$ exceeds $\frac{3}{2}kT$.

Determining the shape of the potential from particle velocity measurements would enable predictions on the existence of a critical temperature T_c, defined as the temperature at which the binding energy of the particles decay below their thermal energy. At T_c, the arrangement of the particles would then change from a regime where convective steering ensures colloidal crystallization to a regime where capillary interactions solely control the arrangement process.

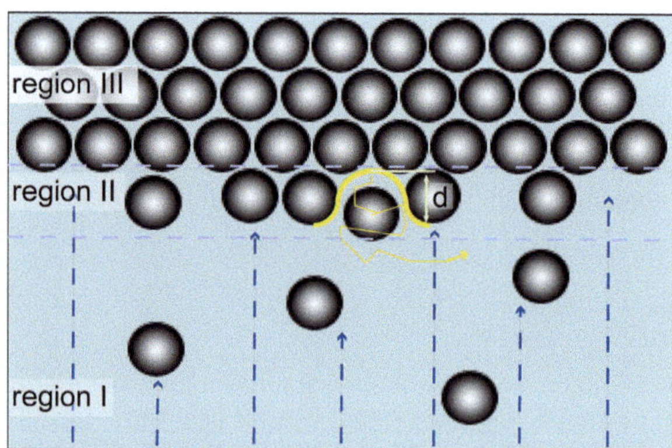

Fig. 3.3 Schematic *top view* of convective particle assembly. In region I the solvent stream advects free particles, in region II the particles arrange due to the force created by the convective stream and the restrictions by neighboring particles, and in region III the deposited particles form a scaffold for the addition of new particles. $d = 2r$ is the distance particles must be able to diffuse against the solvent stream to become unrestricted in lateral motion

3.3 Experimental

All experiments were performed in a climate-controlled laboratory at a temperature of $22 \pm 1\,°C$ and a relative humidity of $40 \pm 5\,\%$.

3.3.1 Particles, Solvents and Substrates

Silicon wafers (Asahi Kasei, Tokyo, Japan, $\langle 100 \rangle$-oriented) were used as substrates. The substrates were cleaned using isopropanol (Sigma-Aldrich, Deisenhofen, Germany, puriss. p. a.) and deionized water (ultrapure quality from a Millipore unit), subsequently, in an ultrasonic bath. Immediately before use, the substrates were hydrophilized by oxygen plasma treatment at a pressure of 0.3 mbar and RF power of 100 W for 5 min in the quartz tube of a commercial plasma reactor (low-pressure reactor PICO, RF source at 13.56 MHz, Diener electronic, Ebhausen, Germany). All substrates were spontaneously wetted by water after this step.

Aqueous suspensions of emulsion-polymerized polystyrene microspheres with specified diameters of 457 ± 10 nm ('PS500') and 110 ± 5 nm ('PS100') were obtained from Polysciences (Polysciences Europe GmbH, Eppelheim, Germany), the solid content was specified as 2.6 and 4 %, respectively. Before deposition, the suspensions were filtered by pressing them through 0.8 μm cellulose acetate syringe filters (Whatman, Dassel, Germany).

Barium titanate powders ('BTO') with mean particle diameter of 300 nm and a relative permittivity of 809 were obtained from Inframat Advanced Materials (Willington, CT, USA). Aqueous suspensions with a solid content of $\approx$0.1 % were obtained from the powders by a route described by LI et al. [41].

Suspensions of 1 μm silica particles ('SiO1000') were prepared according to NOZAWA et al. [42]. Briefly, a solution of 5 ml TEOS ($Si(OC_2H_5)_4$, 98 %, Sigma-Aldrich) in 30 ml ethanol (99.9 %, Eckvos) were added to a solution of 9.5 ml ammonia (NH_4OH, 28 %) in 50 ml of ethanol under constant stirring at 500 rpm. Addition rate was kept constant at 0.05 ml/min.

3.3.2 Experimental Setup, Film Deposition and Video Microscopy

The setup was described in Chap. 2. It was designed for horizontal convective particle assembly in blade-coating like geometry with concurrent observation by light microscopy. It consisted of a temperature-controlled sample holder that was moved by a precision linear stage. The suspension reservoir was formed by a capillary bridge between the substrate and a blade mounted above the substrate. The entire setup was mounted on the stage of conventional light microscope to enable video microscopy.

For film deposition, approximately $100\,\mu l$ of particle suspension were injected into the gap between the blade and the substrate. The suspensions was allowed to equilibrate its temperature with the substrate for 5 min. Depositions were made in the temperature interval from 8 to 40 °C. The withdrawal rate of the substrate was chosen to deposit particle monolayer, which required withdrawal velocities of 10–$20\,\mu m/s$ at room temperature and down to 50 nm/s for depositions below 10 °C. The film growth front remained at a steady position within the field of view of the objective under these conditions.

Videos of the depositions, either of the particle transport on the meniscus side of the growth front or of the particle film side of the growth front, were taken at frame rates of $30\,s^{-1}$ and at roughly 500x magnification (frame size $210 \times 170\,\mu m^2$). Videos consisting of 600 frames (20 s length) were used for evaluation with particle image velocimetry or particle tracking. Prior to evaluation the video frames were adjusted in brightness and contrast using the ImageJ software [43].

3.3.3 Particle Image Velocimetry

A particle image velocimetry algorithm was used to analyze the velocity of particles when suspended in the solvent or embedded in the particle film. The videos were imported into Matlab [44]. Our algorithm selects rectangular interrogation areas and calculates the cross-correlation between each pair of interrogation areas of two subsequent frames. The maxima in correlation are used to estimate the displacement of the particles between the frames. More detailed descriptions of particle image velocimetry by cross-correlation can be found in the literature [45, 46]. We used the equation given by Pust [47] for calculating the cross-correlation coefficient R between the interrogation areas A and B for an offset $[m, n]$:

$$R_{AB}(m, n) = \frac{\sum_i \sum_j [A(i, j) - \overline{A}] \cdot [B(i + m, j + n) - \overline{B}]}{\sqrt{\sum_i \sum_j [A(i, j) - \overline{A}]^2 \cdot \sum_i \sum_j [B(i + m, j + n) - \overline{B}(m, n)]^2}}.$$

$$(3.9)$$

$\overline{A}$ and $\overline{B}$ denote the interrogation area mean values and i and j the pixel indexes.

Averaged velocity profiles for the entire movie were obtained by averaging the displacements obtained for each interrogation area over all frames. The flow profile is symmetrical; we averaged over interrogation areas with the same distance from the film edge to obtain a 1-dimensional flow profile. Figure 3.4 illustrates the evaluation process. Finally, an inverse parabola was fit to the flow profile using the Origin software [48]:

$$v(x) = -\frac{1}{a + b \cdot x + c \cdot x^2}.$$

$$(3.10)$$

Parameter b and c only depend on the setup geometry, which was constant. They were averaged over all measured velocity profiles. An exponential function derived

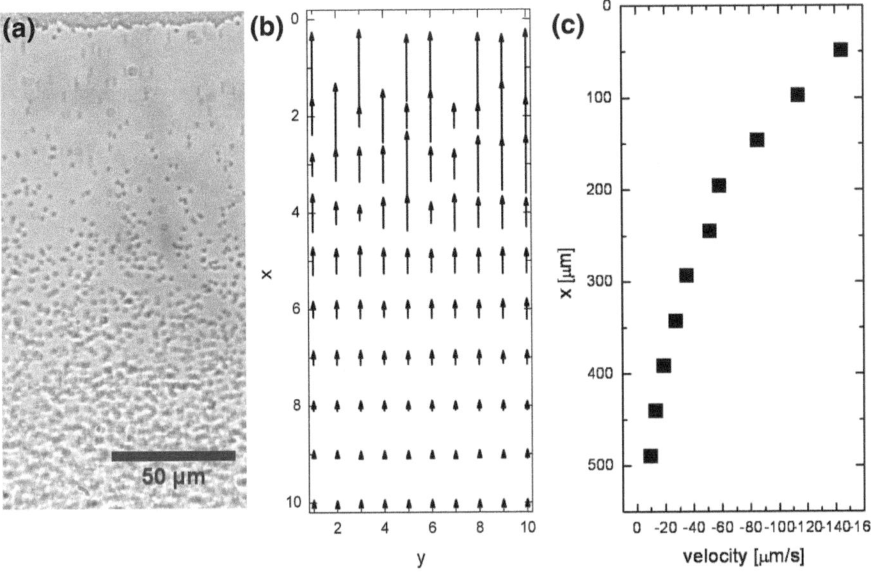

Fig. 3.4 The process of particle image velocimetry: From two subsequent video frames (**a**) shifts in the correlation maximum for each interrogation area are calculated (**b**), which leads to the velocity profile after calibration to the magnification and the frame rate of the video (**c**). The example shows SiO1000 particles at 25 °C

from the law of CLAUSIUS- CLAPEYRON was fit to parameter $a(T)$:

$$a(T) = -\frac{1}{v(0, T)} = -\frac{1}{d \cdot e^{-\frac{e}{T}}}.$$ (3.11)

The isolated particles in moving suspension provided sufficient optical contrast for particle image velocimetry. In the dense packing of the assembled film, however, single particles could not be resolved anymore. We therefore used BTO tracer particles to analyze particle mobility in the films. Their high refractive index causes strong light scattering so that the tracers appear as bright spots in dark field imaging of the particle film during deposition. The size difference between BTO particles and PS100 particles probably causes differences in their motion, but we believe that the tracer particles' motion characterize the overall mobility in the assembled film well.

The evaluation of the tracers' motion followed the scheme described above with two exceptions: to compensate for the much slower dynamics in the particle film, cross-correlations were calculated not for subsequent frames but with twenty frames spacing. The motion of tracer particles does not represent a steady state. We therefore did not average over the entire movies but calculated trajectories of the tracers as they move through the film.

3.3.4 Particle Tracking

A particle tracking algorithm by KOUMOUTSAKOS et al. [49] was used to reconstruct the trajectories of single particles in the liquid volume during deposition. This algorithm tracks the positions of bright spots in image series. The high particle concentrations of convective particle assembly and the two-dimensional imaging of the meniscus make the tracking of individual particles difficult. Lower particle concentrations would require a lower deposition rate, which impairs particle film quality (compare Chap. 2). We therefore used BTO particles as tracers. Particle tracking was performed only for PS500 particles which match the size of the tracer particles.

3.3.5 Film Quality Analysis

The quality of particle films was evaluated by counting the next neighbors of each particle. Electron micrographs were obtained using a FEI Quanta 400F scanning electron microscope (FEI, Europe, Eindhoven, Netherlands). The particle tracking algorithm was used to extract the coordinates and count the individual particles' centers from the micrographs. The particles' projected A_p area was estimated from the number of particles N and the area of a single particle with average diameter: $A_p = N \cdot \pi \cdot \bar{r}^2$. The coverage of the substrate was calculated by dividing the substrate area by the particles' area. Next neighbors for each particle were then counted by counting the number of particle centers within a distance of two radii from the particle's center. Roughly 2500 particles per image were analyzed to obtain the number distribution of next neighbors.

3.4 Results

In the following, we analyze the velocity of particles during their convective assembly using the potential model. We find that the transport potential describes the migration of particles from the bulk to the assembly region well. In the assembly region, depending on temperature, one of the two different assembly mechanisms occurs. Mobility analysis proves that formation of a regular arrangement proceeds via an intermediate arrangement with high mobility in capillary crystallization.

3.4.1 Transport and Binding Potentials

The velocity distribution of particles that are transported from the bulk to the assembly region is strongly temperature-dependent. Figure 3.4 illustrates experimentally measured velocity profiles. Such profiles were obtained for SiO1000-particles at substrate temperatures of 25, 33 and 40 °C and for the PS500 particles at substrate temperatures of 15, 19, 25, 33 and 40 °C. In all cases, the particles were found to accelerate towards the assembly region, as expected.

All particle velocity profiles fit Eq. 3.10 with regression coefficients better than 0.95 (3.5). Transport of particles to the assembly region in our experiments is described by Eq. 3.10 with $b = 1.6 \cdot 10^{-19}$ s/μm^2 (with a standard deviation of $1.9 \cdot 10^{-18}$ s/μm^2) and $c = 9.4 \cdot 10^{-7}$ s/μm^3 (with a standard deviation of $1.4 \cdot 10^{-6}$ s/μm^3) from averaging over all measurements. The variations in the linear and quadratic term of the flow profile indicate that the assumption of a static meniscus shape is critical over the full range of temperatures and withdrawal velocities tested in the experiments. The error in the binding potential calculated below caused by the uncertainties in b and c, however, stays below 0.9 % for all evaluated particle sizes.

All particles reach their maximum velocity when they arrive at the film growth front. Maximal velocities from PIV $v(0, T) = -1/a(T)$ are plotted in Fig. 3.5. They fit the predictions of Eq. 3.11 with a regression coefficient of (0.8) and parameters $d = 1.27 \cdot 10^{10}$ μm/s (standard error of $5.73 \cdot 10^{9}$ μm/s) and $e = 5376$ K (standard error of 1453 K) (errors estimated from the goodness of the fit). As discussed in Sect. 3.5, there may be additional parameters not included in the model contributing to $v(0)$, which limit the goodness of the fit.

We are now in a position to recover the force $F(v, x, r, T)$ that would act on a stagnant particle in the fluid with velocity $v(x, T)$ and construct the corresponding potential. The derived fluid velocity $v(x, T)$ depends only on the distance from the film growth front and the temperature, integration over the drag force $F(v, x, r, T)$

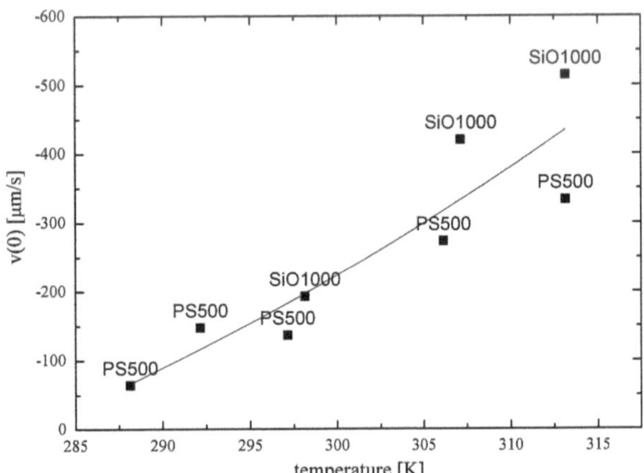

Fig. 3.5 The velocities $v(0, T)$ of the convective fluid stream calculated for $x = 0$ at film growth front. The *solid line* is the result of the fit with Eq. 3.11. The trend follows the predictions by the model, the fluid velocity increases with temperature. The labels indicate the tracer particles used in the measurements

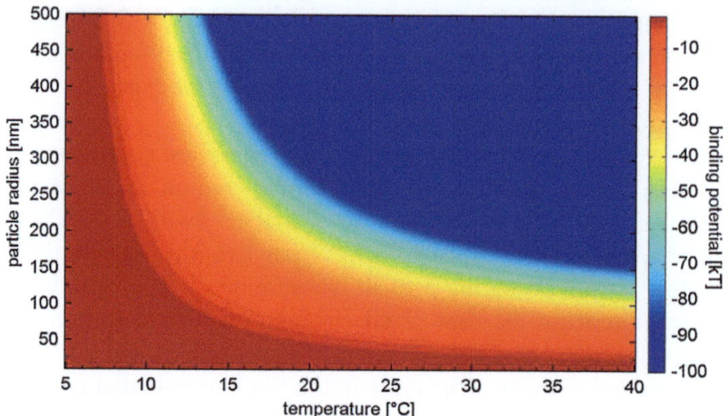

Fig. 3.6 Plot of the depth of the binding potential in units of kT as calculated from Eq. 3.8. The plot is truncated at 100 kT to maintain resolution at lower energies. A distinct dependency on the particle radius and the temperature can be found, below $\approx$250 nm and $\approx$20 °C the potential energy rapidly decays to kT

yields the potential. Here, we are mainly interested in the magnitude of the binding potential, i.e., the potential directly at the film growth front. To ease applicability in the laboratory we give the results of the calculations in units of °C instead of K.

Figure 3.6 is a color-coded map of the binding potential Φ_b as a function of the particle radius and the substrate temperature. The binding potential shows a strong dependency on particle radius and temperature and diverges for large particles and high temperatures, where Φ_b takes values of several hundred kT. Below a particle radius of $\approx$250 nm and below a temperature of $\approx$20 °C, Φ_b rapidly decays to kT. This is insufficient to confine the particles at the film growth front and maintain order there: assembly requires Φ_b exceeding 3/2 kT. Particles have enough thermal energy to leave shallower potential wells and escape from their lattice position.

The effect of particle size is highlighted in Fig. 3.7. As temperature increases, the depth of the potential well rapidly grows beyond 3/2 kT for 500 nm-diameter particles, by more than 50 kT in the range from 5 to 18 °C. In contrast, the depth of the potential well only increases by 2.8 kT between 5 and 20 °C for 100 nm-diameter particles.

Critical temperatures at which the potential well reaches 3/2 kT are shown in Fig. 3.8. Note that only small particles (with radii below 100 nm) exhibit critical temperatures above the dew point. For large particles, critical temperatures are around the dew point which is close to 5 °C for the ambient conditions in the laboratory.

A second important result is the magnitude of the transporting potential (Eq. 3.7). Our model predicts that even small particles are transported and held in the meniscus. For example, a particle with a radius of 5 nm at a substrate temparture of 10 °C still encounters a potential difference of $\approx$50 kT between the film growth front and the reservoir. An accumulation of particles in the meniscus thus will happen at all practical temperatures.

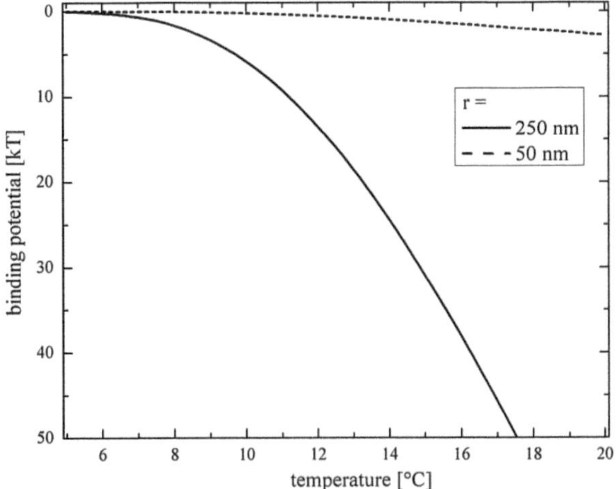

Fig. 3.7 Plot of the depth of the binding potential as a function of the substrate temperature for two particle diameters. In the case of 500 nm-diameter particles the depth of the binding potential well increases by more than 50 kT in the given temperature range, while it increases only by 2.8 kT in the case of 100 nm-diameter particles

Fig. 3.8 Plot of the critical temperature T_c at which the potential well reaches 3/2 kT as a function of the particle radius. For large particles this temperature coincides with the dew point in the laboratory, while it diverges for particles below 100 nm. The inset gives an expanded plot of the sub-100 nm region

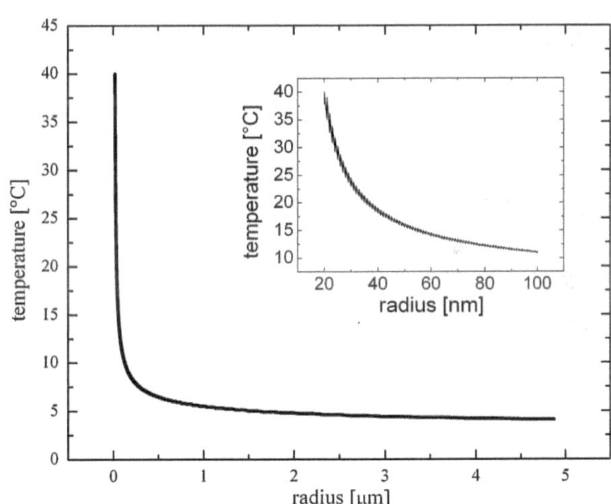

3.4.2 Assembly Mechanisms

Assembly cannot take place if particles are not transported to the assembly region. Crystallization at the film growth front cannot take place if the binding potential well at this position is too shallow. In the following, we analyze the trajectories of particles outside and inside the deposited film at various temperatures to find how they assume regular lattice positions.

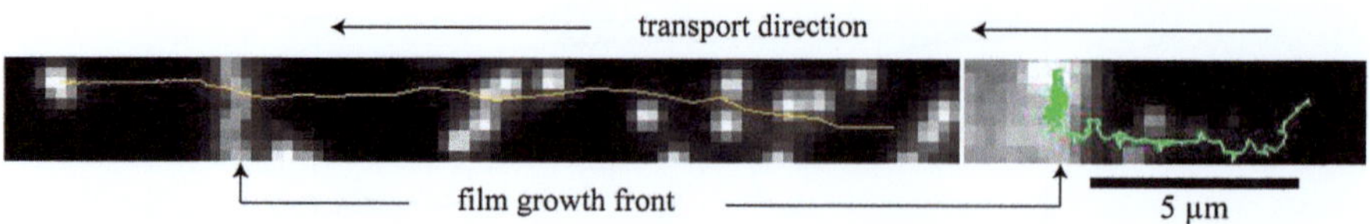

Fig. 3.9 Trajectories of BTO particles co-deposited with PS500 particles at 20 °C (*left*) and at 8 °C (*right*). The convective stream transported the particles from the right to the film growth edge. At 20 °C, the particle is build into the film and follows the withdrawn substrate without lateral displacement, while the particle at 8 °C stays mobile and diffuses along the film growth edge

We used barium titanate (BTO) tracer particles to analyze particle trajectories above and below the critical temperature. Figure 3.9 shows two representative trajectories of BTO particles that were co-deposited with PS500 particles at 20 and 8 °C. For such particles, the binding potential well is too shallow to enforce crystallization below 10 °C; at 8 °C the binding potential Φ_b becomes 0.6 kT.

The particle trajectories at 20 °C hardly deviate from the smooth stream lines of the solvent. They end at the growth front of the particle film, where all independent motion of tracer particles ceases and they merely follows the substrate. In contrast, at 8 °C the transport of the particles is much slower. Trajectories are still directed towards the growth front but jagged by random detours. They do not end at the film growth front: particles diffuse along the front as predicted by the model.

If particles can diffuse along the growth front, convective steering cannot be the prevailing mechanism of crystallization. We find, however, that ordered films still emerge. Capillary forces must be responsible for this crystallization. As KRALCHEVSKY et al. pointed out [32], the immersion forces even of very small particles are stronger than thermal agitation. When the liquid film enclosing the particles' film thins beyond the diameter of the particles at some distance from the growth edge, strong capillary interactions pull the particles into hexagonal close packed structures.

We investigated the motion of particles inside the film by PIV to analyze this second assembly mechanism. A monolayer of PS100 particles was deposited. The critical temperature for 100 nm-diameter particles is sufficiently above the dew point to avoid condensing water. For such particles, the binding potential Φ_b is expected to be 0.7 kT at 13 °C and 2 kT at 17 °C, the temperatures we chose for deposition. Figure 3.10 shows one frame of the dark field videos of the PS100 particle film with BTA tracer particles. It is easy to observe mobility inside the film by observing the BTA particles' displacements. Figure 3.11 indicates directions and magnitudes of the tracer displacement from one video frame to another with 0.6 s delay. Little mobility is seen at 17 °C. The particles merely follow the substrate. The particle film at 13 °C exhibits much greater mobility, in particular at the first 100 μm behind the film growth front. Random rearrangement of the particles occurs directly behind the front, while the particles further away from the front increasingly follow the withdrawn substrate. Root mean square displacements shown in Fig. 3.12 quantify this trend. The displacements were obtained by averaging over all interrogation areas with the

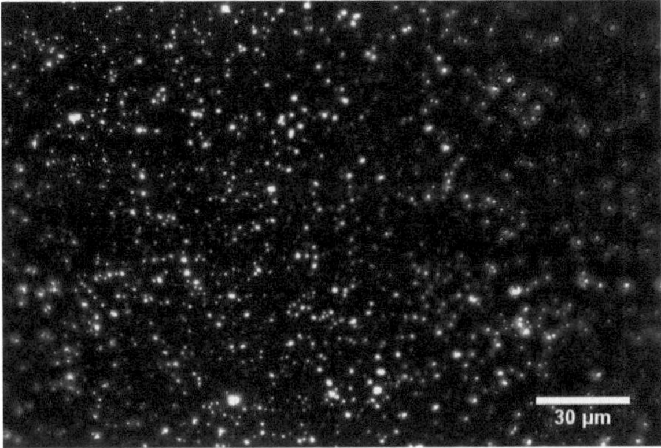

Fig. 3.10 Video frame showing the deposition of a 100 nm polystyrene particles and BTO particles. The BTO particles aid the observation of mobility in the particle films by their high contrast in refractive indexes

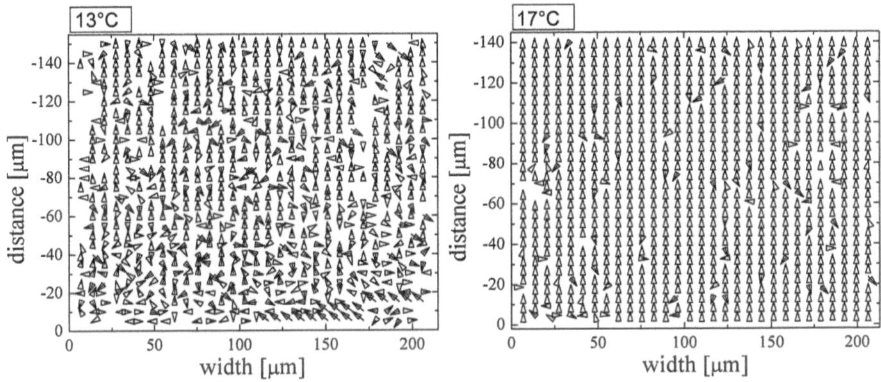

Fig. 3.11 Displacements between two video frames with 0.6 s lag observed in 100 nm-diameter polystyrene particle films with BTO tracer particles at 13 °C and at 17 °C. The particles in the film deposited at 13 °C exhibit especially in the first 100 µm from the film growth front random displacements, whereas the particles in the film deposited at 17 °C merely follow the withdrawn substrate

same distance from the film growth edge and correcting for the steady withdrawal of the substrate.

The observed displacements probably indicate the effect of rearrangement by capillary interactions. In the following we show that the assembled films have characteristic features that indicate their assembly mechanism.

Films of PS100 particles without the addition of BTO tracer particles were deposited at 13 and 17 °C. Figure 3.13 shows electron micrographs of the resulting particle arrangements. Both exhibit hexagonal dense packing of the polystyrene spheres. The surface coverage was 69.4 % for the film deposited at 13 °C and 74.3 % for the film deposited at 17 °C, both similar and far below the theoretical maximum of

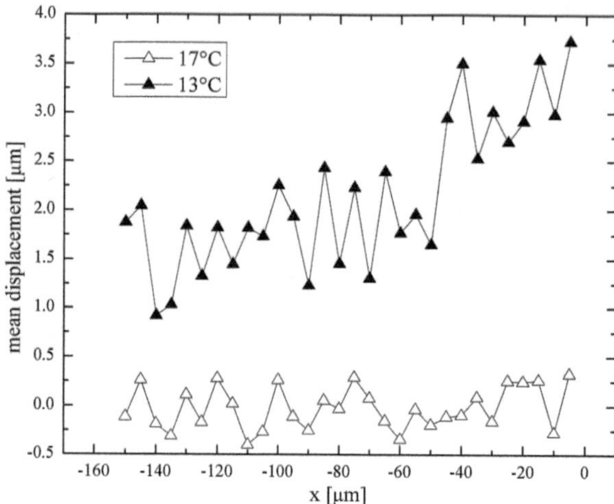

Fig. 3.12 Averaged root mean square values of the displacements within 100 nm particle films deposited at 13 and 17 °C. The particles in the film deposited at 13 °C exhibit high initial average displacements, while the film deposited at 17 °C keeps a static structure. At 17 °C the particles are held in position by the convective stream

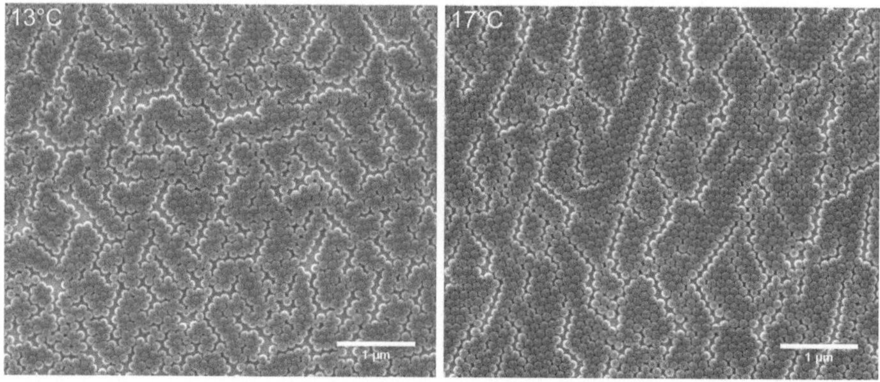

Fig. 3.13 Electron micrographs of 100 nm particle films deposited at 13 °C (*left*) and 17 °C (*right*)

90.7 %. Pronounced differences in the particle packing between the two depositions are quantified by next-neighbors-analysis (Fig. 3.14). While most particles deposited at 17 °C have 6 next neighbors, most particles deposited at 13 °C have only 4 next neighbors. At 17 °C, most particles have a complete hexagonal neighborhood, while at 13 °C, most particles sit at edges of hexagonal grains. The differences in assembly mechanism above and below T_c lead to different particle packings in the deposited film. Below the critical temperature, small-grained particle films are formed.

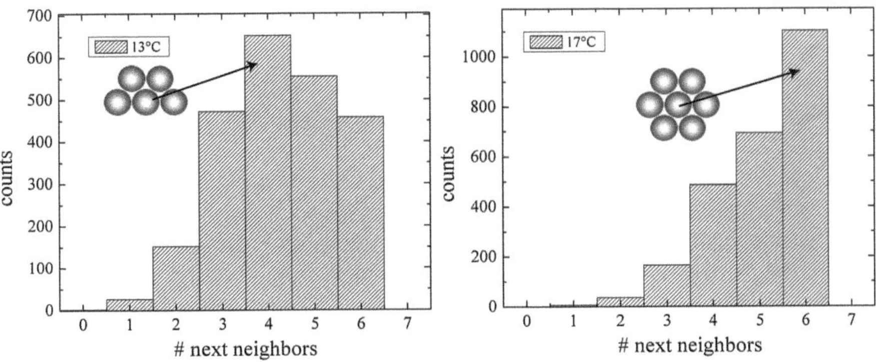

Fig. 3.14 Distribution of the numbers of next neighbors for 100 nm particle films deposited at 13 and 17 °C. Most particles have an incomplete neighborhood at 13 °C, thus sit at edges of crystallites, while at 17 °C most particles have a complete neighborhood and are situated within crystalline packings

3.5 Discussion

Convective flows can fulfill two tasks in convective particle assembly: first, they transport particles from the reservoir to the assembly region. Second, they can assemble the particles into regular lattices, but only at high evaporation rates.

The simplified model of the flow derived above immediately suggests that

1. convective flows efficiently transport and assemble particles larger than 500 nm under all realistic conditions,
2. particles below 500 nm are efficiently transported to the assembly region, but the convective flow is only strong enough to assemble them into regular packings above a critical temperature T_c that is higher than the dew point of the solvent.

This behavior is reflected in the trajectories of single particles and the final packing. We find that

1. for $T > T_c$, particles are transported to the film growth front, attached to it and are immobilized,
2. for $T < T_c$, particles are transported to the film growth edge, but merely diffuse along it.
3. for $T < T_c$, particles remain mobile even in the deposited film, with decreasing mobility with the distance from the growth edge,
4. above and below T_c, the dried film exhibits hexagonal close-packing,
5. for $T < T_c$, deposition leads to films with smaller crystalline grains.

The results indicate a transition of the assembly mechanism around T_c:

Above T_c, assembly proceeds mainly in two distinctly different regions. Region III, the wet particle film, is static with no particle rearrangement. Region II only represents the boundary between freely suspended particles and the static particle film.

All dynamics of the assembly process take place in region I, the liquid meniscus. In this region, particles are transported from the bulk to the growth front and distributed into the available niches. The previously deposited particle film forms a static scaffold that modulates the solvent flow and thus defines the trajectory of incoming particles.

Below T_c, the regions are less clearly delineated. In region I particles are transported towards the film growth front, but their motion is only weakly constrained when they arrive there. In region II, the particles constantly leave their binding sites and rearrange. A static scaffold as required for convective steering cannot form and the particle film exhibits a random, loose structure. This random arrangement crystallizes in Region III due to capillary bridges.

This capillary-induced crystallization probably starts at several nucleations sites with approximately uniform distance from the growth front, a mechanism that finally leads to the fine-grained structure of the packing we observe. The random precursor film in region II is less dense than the hexagonal packing and the capillary attraction produces cracks in the particle film. However, further particles can be added during the crystallization due to the high mobility in region II. Consequently, capillary-crystallized films exhibit more cracks than those assembled by convective steering but only little less density.

The simple model introduced here is easily extended. We discuss some limitations in the following.

Presently, in the model the particle radius only is considered when calculating the Stokes' drag. Experimentally we find size-dependent fluid velocities $v(0)$ (Fig. 3.5) particularly at high temperatures. Such deviations can originate from particle size effects on the meniscus geometry and on the evaporation rate of the drying particle film. The latter may also depend on the exact contact angle on the particles and thus, the particle material, just as the pressure gradient generated in the wet particle film.

Interactions between particles beyond hard exclusion are entirely neglected in our model. The repulsive interactions of the charge-stabilized particles result in a finite distance between the particles in the wet deposited films even above T_c. They are overcome only in the last stages of drying, which causes the observed drying cracks. Possibly, the distance between the assembled particles can be derived from the magnitude of the binding potential and the cracking be predicted.

In summary, we believe our model to be a good approximation. Care must be taken when interpreting the data for small particles. Further measurements of $v(x)$ with particles below 100 nm in diameter, possibly using fluorescent labels, could reveal the exact shape of $T_c(r)$. The potential introduced here is a convenient parameter to plot against properties of the film such as surface coverage or long-range order. If necessary, it is easily extended to cover the gravitation potential in vertical depositions or additional frictional terms due to particle-particle or particle-substrate interactions.

3.6 Conclusion

Convective particle assembly was described with a potential energy model. The model based on the flow profile of the convective flow and the drag acting on the particles. By applying microscopic particle image velocimetry a parabolic flow profile and an exponential temperature dependency of the fluid velocity was determined. From the temperature dependency a transition temperature was predicted, below which the convective flow does not confine the particles anymore at the film growth front. The trajectories of traces particles deposited below and above the predicted transition temperature proved a failing confining and a diffusion of particles along the film growth front.

Hexagonal crystalline particle films formed in either case, albeit with different surface coverages. The transition temperature thus marks the transition between two assembly mechanisms, convective steering and capillary crystallization. Convective steering assembles free particles to a static particle film. Capillary crystallization assembles the particles after they have been deposited in a sparse, dynamic film region with high particle mobility.

Hexagonal crystallinity thus can be expected to always be formed from well-stabilized, monodisperse particle suspensions in a stable convective process. Isotropic, amorphous particle coatings could never form because one of the two assembly mechanisms ensures crystallization at every temperature that leads to film deposition at all. Film structures ranging from ordered monocrystalline films to small-grained percolating coatings with low surface coverages are accessible, however, by adjusting the deposition conditions. In future experiments, amorphous coatings may be introduced by additional quenching of the assembly process at a defined stage. Quenching could be achieved by tailored particle-surface interactions, particle-particle interactions or switchable external fields.

References

1. H. Zheng, M.C. Berg, M.F. Rubner, P.T. Hammond, Controlling cell attachment selectively onto biological polymer-colloid templates using polymer-on-polymer stamping. Langmuir **20**, 7215–7222 (2004)
2. C.S. Chen, M. Mrksich, S. Huang, G.M. Whitesides, D.E. Ingber, Geometric control of cell life and death. Science **276**, 1425 (1997)
3. D.M. Glenn, G.J. Puterka, T. van der Zwet, R.E. Byers, C. Feldhake, Hydrophobic particle films: a new paradigm for suppression of arthropod pests and plant diseases. J. Econ. Entomol. **92**, 759–771 (1999)
4. W. Ming, D. Wu, van Benthem, G. de With, Superhydrophobic films from raspberry-like particles. NANOletters **5**, 2298–2301 (2005)
5. B.G. Prevo, E.W. Hon, O.D. Velev, Assembly and characterization of colloid-based antireflective coatings on multicrystalline silicon solar cells. J. Mater. Chem. **17**, 791–799 (2007)
6. P. Krone, C. Brombacher, D. Makarov, K. Lenz, D. Ball, F. Springer, H. Rohrmann, J. Fassbender, M. Albrecht, Nanocap arrays of granular CoCrPt:SiO2 films on silica particles: tailoring of the magnetic properties by Co+ irradiation. Nanotechnology **21**, 385703 (2010)

7. H.J. Nam, D.-Y. Jung, G.-R. Yi, H. Choi, Close-packed hemispherical microlens array from two-dimensional ordered polymeric microspheres. Langmuir **22**, 7358–7363 (2006)
8. P. Kumnorkaew, Y.-K. Ee, N. Tansu, J.F. Gilchrist, Investigation of the deposition of microsphere monolayers for fabrication of microlens arrays. Langmuir **24**, 12150–12157 (2008)
9. Y. Zhao, I. Avrutsky, B. Li, Optical coupling between monocrystalline colloidal crystals and a planar waveguide. Appl. Phys. Lett. **75**, 3596 (1999)
10. A.S. Dimitrov, K. Nagayama, Continuous convective assembling of fine particles into two-dimensional arrays on solid surfaces. Langmuir **12**, 1303–1311 (1996)
11. T. Still, W. Cheng, M. Retsch, R. Sainidou, J. Wang, U. Jonas, N. Stefanou, G. Fytas, Simultaneous occurrence of structure-directed and particle-resonance-induced phononic gaps in colloidal films. Phys. Rev. Lett. **100**, 194301 (2008)
12. Y. Xu, B. Zhang, W.H. Fan, D. Wu, Y.H. Sun, Sol-gel broadband anti-reflective single-layer silica films with high laser damage threshold. Thin Solid Films **440**, 180–183 (2003)
13. Y. Xia, N.J. Halas, Shape-controlled synthesis and surface plasmonic properties of metallic nanostructures. MRS Bull. **30**, 338 (2005)
14. A.D. Ormonde, E.C.M. Hicks, J. Castillo, R.P.V. Duyne, Nanosphere lithography: fabrication of large-area Ag nanoparticle arrays by convective self-assembly and their characterization by scanning UV-Visible extinction spectroscopy. Langmuir **20**, 6927–6931 (2004)
15. N.H. Finkel, B.G. Prevo, O.D. Velev, L. He, Ordered silicon nanocavity arrays in surface-assisted desorption/ionization mass spectrometry. Anal. Chem. **77**, 1088–1095 (2005)
16. N.D. Denkov, O.D. Velev, P.A. Kralchevsky, I.B. Ivanov, H. Yoshimura, K. Nagayama, Mechnism of formation of two-dimensional crystals from latex particles on substrates. Langmuir **8**, 3183–3190 (1992)
17. P. Jiang, J.F. Bertone, K.S. Hwang, V.L. Colvin, Single-crystal colloidal multilayers of controlled thickness. Chem. Mater. **11**, 2132–2140 (1999)
18. H.J. Schöpe, A.B. Fontecha, H. König, J.M. Hueso, R. Biehl, Fast microscopic method for large scale determination of structure, morphology, and quality of thin colloidal crystals. Langmuir **22**, 1828–1838 (2006)
19. B.G. Prevo, D.M. Kuncicky, O.D. Velev, Engineered deposition of coatings from nano- and micro-particles: a brief review of convective assembly at high volume fraction. Colloids Surf. A **311**, 2–10 (2007)
20. L. Malaquin, T. Kraus, H. Schmid, E. Delamarche, H. Wolf, Controlled particle placement through convective and capillary assembly. Langmuir **23**, 11513–11521 (2007)
21. E. Vekris, V. Kitaev, D.D. Perovic, J.S. Aitchinson, G.A. Ozin, Visualization of stacking faults and their formation in colloidal photonic crystal films. Adv. Mater. **20**, 1110–1116 (2008)
22. G.S. Lozano, L.A. Dorado, R.A. Depine, H. Miguez, Towards a full understanding of the growth dynamics and optical response of self-assembled photonic colloidal crystal films. J. Mater. Chem. **19**, 185–190 (2009)
23. K. Chen, S.V. Stoianov, J. Bangerter, H.D. Robinson, Restricted meniscus convective self-assembly. J. Colloid Interface Sci. **344**, 315–320 (2010)
24. E.C.H. Ng, K.M. Chin, C.C. Wong, Controlling inplane orientation of a monolayer colloidal crystal by meniscus pinning. Langmuir **27**, 2244–2249 (2011)
25. R.D. Deegan, O. Bakajin, T.F. Dupont, G. Huber, S.R. Nagel, T.A. Witten, Capillary flows as the cause of ring stains from dried liquid drops. Nature **389**, 827–829 (1997)
26. R.D. Deegan, O. Bakajin, T.F. Dupont, G. Huber, S.R. Nagel, T. Witten, Contact line deposits in an evaporating drop. Phys. Rev. E: Stat., Nonlin, Soft Matter Phys. **62**(1), 756–765 (2000)
27. D.J. Norris, E.G. Arlinghaus, L. Meng, R. Heiny, L.E. Scriven, Opaline photonic crystals: How does self-assembly work? Adv. Mater. **16**, 1393–1399 (2004)
28. D. Gasperino, I. Meng, D.J. Norris, J.J. Derby, The role of fluid flow and convective steering during the assembly of colloidal crystals. J. Cryst. Growth **310**, 131–139 (2008)
29. D.D. Brewer, J. Allen, M.R. Miller, J.M. de Santos, S. Kumar, D.J. Norris, M. Tsapatsis, L.E. Scriven, Mechanistic principles of colloidal crystal growth by evaporation-induced convective steering. Langmuir **24**, 13683–13693 (2008)

30. K. Nagayama, Fabrication of two-dimensional colloidal arrays. Phase Transitions **45**, 185–203 (1993)
31. N.D. Denkov, O.D. Velev, P.A. Kralchevsky, I.B. Ivanov, H. Yoshimura, K. Nagayama, Two-dimensional crystallization. Nature **361**, 26 (1993)
32. P.A. Kralchevsky, K. Nagayama, *Particles at Fluid Interfaces and Membranes* (Elsevier, Amsterdam, 2001)
33. D.R. Line (ed.), *CRC Handbook of Chemistry and Physics* (CRC Press, New York, 2009)
34. J. Kleinert, S. Kim, O.D. Velev, Electric-field-assisted convective assembly of colloidal crystal coatings. Langmuir **26**, 10380–10385 (2010)
35. P. Born, S. Blum, A. Munoz, T. Kraus, Role of the meniscus shape in large-area convective particle assembly. Langmuir **27**, 8621–8633 (2011)
36. A.S. Dimitrov, T. Miwa, K. Nagayama, A comparison between the optical properties of amorphous and crystalline monolayers of silica particles. Langmuir **15**, 5257–5264 (1999)
37. M.E. Calvo, O.S. Sobrado, G. Lozano, H. Miguez, Molding with nanoparticle-based one-dimensional photonic crystals: a route to flexible and transferable Bragg mirrors of high dielectric constant. J. Mater. Chem. **19**, 3144–3148 (2009)
38. M. Bunzendahl, P.L.-V. Schaick, J.F.T. Conroy, C.E. Daitch, P.M. Norris, Convective self-assembly of stoeber sphere arrays for syntactic interlayer dielectrics. Colloids Surf., A **182**, 275–283 (2001)
39. I.N. Bronstein, K. Semendjajew, G. Musiol, H. Mühlig, *Taschenbuch der Mathematik* (Verlag Harri Deutsch, Frankfurt, 2001)
40. A. von Blaaderen, Colloids under external control. MRS Bull. **2004**, 85–90 (2004)
41. C.-C. Li, S.-J. Changb, J.-T. Leec, W.-S. Liao, Efficient hydroxylation of BaTiO3 nanoparticles by using hydrogen peroxide. Colloids Surf., A **361**, 143–149 (2010)
42. K. Nozawa, H. Gailhanou, L. Raison, P. Panizza, H. Ushiki, E. Sellier, J.P. Delville, M.H. Delville, Smart control of monodisperse Stöber silica particles: effect of reactant addition rate on growth process. Langmuir **21**, 1516–1523 (2005)
43. ImageJ. http://rsb.info.nih.gov/ij
44. Matlab. http://www.mathworks.com
45. R.J. Adrian, Particle-imaging techniques for experimental fluid mechanics. Annu. Rev. Fluid Mech. **23**, 261–304 (1991)
46. A.K. Prasad, Particle image velocimetry. Curr. Sci. **79**, 51 (2000)
47. O. Pust, Direct cross-correlation compared with FFT-based cross-correlation, in *Proceedings of the 10th International Symposium on Applications of Laser Techniques to Fluid Mechanics*, Lisbon, Portugal, 2000
48. Origin. http://www.originlab.com
49. I.F. Sbalzarini, P. Koumoutsakos, Feature point tracking and trajectory analysis for video imaging in cell biology. J. Struct. Biol. **151**, 182–195 (2005)

Part II
Temperature-Induced Particle Assembly

Chapter 4
Growth Kinetics in Temperature-Induced Agglomeration

Abstract The short ligand chain length and the strong VAN DER WAALS-attraction of the cores among alkyl thiol-stabilized metal nanoparticles bring the question about whether the interaction among the particles is core- or ligand-dominated. The details of the interaction determine the agglomeration kinetics and therefore the assembly behavior of the particles. In this chapter, the interaction and agglomeration kinetics of monodisperse alkyl thiol-stabilized gold nanoparticles are investigated. A ligand-dominated interaction among the particles and a temperature-sensitive stability of the suspension is found. We find that agglomeration can be induced by cooling the suspension. The agglomeration kinetics of the suspension follows diffusion-limited or reaction-limited mechanisms. The kinetics control the morphology of the formed agglomerates, offering a tool to design the agglomerates. Surprisingly, crystallization is fully suppressed in temperature-induced agglomeration.

4.1 Introduction

The assembly of colloidal particles into regular particle packings, so-called colloidal crystals, has attracted attention because of collective optical and electronic properties [1–10]. Convenient suspensions for self-assembly colloidal crystallization experiments are alkyl thiol-stabilized metal nanoparticles suspended in unpolar solvents. They can be synthesized in large quantities with low size dispersity [11]. The self-assembly usually occurs during agglomeration of the particles in precipitation [9, 12–14] or evaporation [15–19] experiments. The formed agglomerates are investigated after agglomeration in terms of packing structure and morphology.

However, the agglomeration kinetics of alkyl thiol-stabilized nanoparticles are surprisingly poorly investigated. Agglomeration kinetics are known to be a structure directing parameter. In atomic or molecular crystallization the kinetics are known to generate inspiring and complex patterns like the unlimited number of snowflake geometries [20]. In charge-stabilized particle suspensions depending on the kinetics

P. G. Born, *Crystallization of Nanoscaled Colloids*, Springer Theses, DOI: 10.1007/978-3-319-00230-9_4, © Springer International Publishing Switzerland 2013

dense or ramified structures or fractal or regular morphologies are formed in a predictable manner [21–25]. An influence of agglomeration kinetics on the quality of formed colloidal crystals from sterically stabilized nanoparticles has also already been observed [9, 12].

Agglomeration kinetics can be sorted into two main regimes: diffusion-limited agglomeration (DLA) and reaction-limited agglomeration (RLA) [24]. The first category, also termed fast, rapid, BROWNIAN or SMOLUCHOWSKIAN agglomeration, is characterized by a fast agglomeration process where every particle collision leads to aggregation. The latter category, also termed slow agglomeration, is characterized by lower agglomeration rates and a finite probability of aggregation upon particle collision.

The differences between the two regimes are well-known for charge-stabilized suspensions. Charge-stabilized suspensions are intrinsically unstable, the repulsive potential barrier created by the charged surfaces merely provides a kinetic stabilization of the suspensions [24]. The height of the potential barrier defines the agglomeration kinetics of the suspension. If the potential barrier is lower than the thermal energy of the particles, every collision between particles will lead to agglomeration. The suspension will rapidly flocculate. For potential barriers higher than the thermal energy of the particles only a finite chance for particles to overcome the barrier upon collision exists. The probability is given by the BOLTZMANN factor $e^{-\frac{V_{max}}{kT}}$, where V_{max} denotes the height of the potential barrier, k BOLTZMANN's constant and T the absolute temperature.

The differences between the two regimes can be quantified using rate constants and stability ratios. In the early stage of agglomeration the depletion rate of free particles by collision of primary particles can be expressed as

$$\frac{dn}{dt} = -k_{11} \cdot n^2, \tag{4.1}$$

where n is the concentration of primary particles and k_{11} is the rate constant for doublet formation [26]. Following SMOLUCHOWSKI'S seminal theory on agglomeration, the rate constant for diffusion-limited agglomeration can be calculated as

$$k_{11}^{Smol} = \frac{8kT}{3\eta}, \tag{4.2}$$

where η denotes the solvent viscosity [27]. In reaction-limited agglomeration, the depletion rate is reduced due to the finite chance of overcoming the potential barrier, thus $k_{11} \leq k_{11}^{Smol}$. The stability ratio W, defined as the ratio between the maximal rate constant and a measured rate constant, contains information about the height of the potential barrier V_{max},

$$W = \frac{k_{11}^{Smol}}{k_{11}} \approx \frac{2}{\kappa r} e^{\frac{V_{max}}{kT}} \tag{4.3}$$

where κ gives the thickness of the electrical double layer around the particles with radius r [27]. The stability ratio is a convenient tool to determine the kinetics. In a diffusion-limited regime the temperature dependency cancels out and the stability ratio turns constant. In reaction-limited conditions the stability ratio depends strongly on parameters like temperature, pH and ion concentration. A plot of W against one of these parameters maps the agglomeration behavior of the suspension.

In the later stages of agglomeration coalescence of agglomerates dominate the agglomeration rates. This implies a declining rate in diffusion-limited agglomeration as the growing agglomerates continuously diffuse slower [28, 29]. In reaction-limited agglomeration, the agglomerate-agglomerate coalescence probability increases with growing agglomerates due to the higher number of points of contact and the agglomeration rate increases [30, 31]. The reaction-limited agglomeration process therefore slows down only in its final stage, when the slow diffusion of large agglomerates limits the processes more than the finite coalescence probability: the agglomerations becomes diffusion-limited [30].

Sterically stabilized suspensions are usually described by their flocculation point rather than by their agglomeration kinetics. The stability arises from the ligand-solvent interactions, which can be described by a θ-temperature. Above the θ-temperature of a solvent-polymer combination, the polymer chains stretch out, while below the respective θ-temperature the chains collapse. For good solvents above the θ-temperature the ligands thus provide stabilization. Lowering the quality of the solvent or lowering the temperature below the θ-temperature will turn the interaction between the ligand chains attractive and cause agglomeration [24]. The dispersion interactions between the particle cores are typically assumed to contribute only marginally to the interparticle interactions [32–34].

However, for short ligand chains and particle materials with high HAMAKER coefficients this assumption may break down. The steric repulsion generated decays quickly when the surface separation d reaches values comparable to the chain length L. According to DOLAN and EDWARDS [35] the repulsive interaction potential V_{steric} created by end-grafted polymers decays as

$$V_{steric} = 2\frac{N_p\,\mathrm{kT}}{A_s}\,e^{-\frac{3d^2}{2lL}}, \tag{4.4}$$

where N_p/A_s is the number of polymer chains per surface area and l is the length of a linker segment of the polymer chain, for alkanes ≈ 0.11 nm [36]. The rapidly decaying repulsive potential will have an even softer appearance for small nanoparticles due the high surface curvature leading to a radially decreasing ligand number density.

The range of the VAN DER WAALS-interaction potential V_{vdW} can be calculated using the HAMAKER form [37],

$$V_{vdW} = -\frac{1}{3}\,A\left(\frac{r^2}{s^2-4r^2}+\frac{r^2}{s^2}+\frac{3}{2}\,\ln\left(1-4\frac{r^2}{s^2}\right)\right), \tag{4.5}$$

where $s = 2r + d$ is the center-to-center distance of particles with radius r and A the HAMAKER coefficient. With $A \approx 50$ kT for gold particles [38] and a particle radius of 4 nm the dispersion interaction decays only at a surface separation d of ≈ 2.04 to $-3/2$ kT. This corresponds to the thickness of a self-assembled monolayer of tetradecane chains [36].

The exact shape of the total potential certainly depends on the exact value of A, for which literature values differ from 25 to 100 kT [37, 38], on the actual value of the polymer density n_p, and on the surface curvature of the particles. It is thus not entirely clear how the two interactions, core–core and ligand–ligand, contribute to the total interaction potential for small metal nanoparticles with high HAMAKER coefficient stabilized by alkane chains with 10–20 segments (decane to eicosan). Detailed investigations on stearyl alcohol-stabilized silica particle suspensions have shown the ligand-ligand interactions to dominate agglomeration [39–44]. The onset of flocculation has been shown to correlate with a structural transition in the ligand shell [34, 45]. An universal feature of the formed agglomerates is the glass-like arrest in the particle packing [46]. However, silica exhibits an at least one order of magnitude lower HAMAKER coefficient than gold, thus gold nanoparticle suspensions may still exhibit a significant core contribution to the interparticle attraction.

The difference between a core-dominated suspension and a ligand-dominated suspension is mainly the dependency of the interaction on temperature and ligand length. The core-dominated suspension will exhibit a temperature-independent dispersion attraction and thus possibly agglomeration even above the θ-temperature of the solvent-ligand chain combination, and longer ligand chain length L will provide better stability (compare Eq. 4.4). A core-dominated suspension thus will behave like a VAN DER WAALS-gas. In a ligand-dominated suspensions the particles are purely repulsive above the θ-temperature of the solvent-ligand pair, and turn attractive below the θ-temperature. A ligand-dominated suspension is comparable to a polymer solution as described by the FLORY- HUGGINS-theory.

In either case, agglomeration in sterically stabilized suspensions is driven by attractive interaction potentials without a repulsive barrier [35]. The suspensions can reach thermal equilibrium. From a stable state the suspension can be cooled into a metastable or binodal state with finite chances of forming stable and growing clusters. Further cooling leads to an unstable or spinodal state where growth of agglomerates is only limited by diffusion [43]. The agglomeration rates of sterically stabilized suspensions thus exhibit a temperature dependency opposite to that of charge-stabilized suspensions.

The unclear particle interactions in small sterically stabilized metal particles hinders predictive optimization of colloidal crystallization processes and the purposeful use of kinetics to control the morphology of formed agglomerates. We therefore seek to clarify the particle-particle interaction mechanisms. Using transmission electron microscopy, dynamic light scattering and small-angle X-ray scattering model alkyl thiol-stabilized gold nanoparticles were characterized, agglomeration temperatures were determined using dynamic light scattering and from small-angle X-ray scattering interaction potentials of the suspended nanoparticles could be gained. The results indicate a prevalence of ligand-ligand interactions in the evaluated range of ligand lengths. Using this results the kinetics of the thermally induced agglomeration could be recorded with dynamic light scattering. The found kinetics follow the growth laws reported for charge-stabilized suspensions. Finally, the kinetics are compared to the produced agglomerate morphologies. A distinct difference in morphology between agglomerates grown in diffusion-limited processes and reaction-limited processes was observed. Particularly interesting for particle self-assembly is the absence of order among the particle packing despite the low size dispersity of the primary particles and the huge variation of agglomeration kinetics among the different experiments.

4.2 Experimental

The experiments aimed at clarifying the dominant interaction among the particles, the agglomeration kinetics and the structure formation in the suspensions. Therefore three gold nanoparticle suspensions with varying ligand chain length were synthesized, characterized, and destabilized by cooling. The dependency of the agglomeration kinetics on the temperature was tested. Finally, the agglomeration kinetics are compared to the structures formed in the suspension.

4.2.1 Particles and Solvents

A synthetic route adapted from ZHENG et al. [11] was used in this work. Gold nanoparticles were formed in a one-pot reduction of a gold source by an amine–borane complex in the presence of an alkyl thiol. In a typical synthesis, 0.31 g chlorotriphenylphosphine gold (AuPPh$_3$Cl, ABCR 98 %) was stirred in 50 ml benzene (Riedel-de-Hahn >99.5 %) forming a colorless solution. A mixture of 0.53 g tert-butylamineborane (Fluka 97 %) and 0.34 g hexadecyl thiol (Fluka >98 %) was added to the AuPPh$_3$Cl solution and reacted at 55 °C for 2 h. Following completion of the reduction reaction the deep red solution was cooled to room temperature, precipitated by the addition of ethanol and washed by centrifugation and subsequent resuspension in toluene. Finally, the particles were resuspended in heptane.

Gold nanoparticles with dodecyl thiol (C12), hexadecyl thiol (C16) and octadecyl thiol (C18) ligands in heptane were produced by using the respective thiol during the synthesis.

4.2.2 Particle Characterization

The gold concentration of the suspensions was measured by inductively coupled plasma-atomic emission spectroscopy (ICP-AES) on a Horiba Jobin Yvon Ultima 2 instrument. Results from diluted suspensions of the gold colloid were compared to a standard gold salt solutions concentration curve for quantitative analysis.

Information about the ligand shell of the particles was obtained by infrared-spectroscopy (IR). A Bruker Tensor 27 FT-IR spectrometer and a thermostated sample cell with 0.2 mm path length and CaF_2 windows were used. For IR measurements the suspensions were dried and resuspended in deuterated chloroform (d-chloroform). The absorption was determined by measuring against pure d-chloroform as a standard.

The size of the metal core of the particles was measured using transmission electron microscopy (TEM) and small-angle X-ray scattering (SAXS).

TEM was performed using a Philips CM200 TEM operating at 200 keV with a point-to-point resolution of 0.24 nm and a lattice resolution of 0.14 nm. For particle characterization, 10 μl suspension was dried on a copper TEM grid with amorphous carbon coating. To determine the mean size and size distribution of the gold cores of the particles, approximately 400 particles were analyzed per sample using the ImageJ-software [47].

SAXS measurements were performed at the beamline BW4 of the DORIS III synchrotron at DESY, Hamburg, using radiation of a wavelength $\lambda = 0.138$ nm. The suspensions were measured in glass capillaries with nominal 2 mm outer diameter (Hilgenberg). Before measuring, all samples were filtered by pressing them through 0.2 μm PTFE syringe filters (Whatman). The capillaries were sealed with glue after filling to prevent evaporation of the solvent. The samples were placed in a thermostated sample holder connected to an external water recirculator (Lauda-Königshofen) with a temperature stability better than $\pm 0.5\,°C$. Prior to measurements, the samples were homogenized at elevated temperatures (35–40 °C). The sample-detector distance was set to cover q-ranges from $1.3 \cdot 10^{-1}$ to $3.5\,nm^{-1}$ with the marCCD165 detector (Marresearch). The scattered x-rays were integrated over 10 min. The obtained two-dimensional intensity patterns were azimuthally averaged to obtain the one-dimensional $I(q)$ profile. The curves were subsequently normalized using the synchrotron ring current and background corrected by subtraction of pure solvent measurements. Size and size dispersity of the particle metal cores were determined by fitting a polydisperse particle form factor to $I(q)$ profiles of dilute samples. The details of the SAXS data evaluation can be found in the appendix of this chapter.

The effective particle interaction potential was derived from the structure factor $S(q)$ of the suspensions measured with SAXS. $S(q)$ was obtained by dividing the $I(q)$ curves by a scaled curve obtained from a dilute sample. $S(q)$ was determined in 2 °C-steps from 30 to 10 °C in a q range from $1.3 \cdot 10^{-1}$ to $3.5\,nm^{-1}$. The radial pair distribution function $g(s)$ was derived by FOURIER-transform of $S(q)$. The depth of the effective pair interaction potential V_{min} was estimated by applying $g(s) \approx e^{V(s)/k_B T}$ [42].

The hydrodynamic radius of the samples was measured using dynamic light scattering (DLS). DLS analysis was performed using a Wyatt Technology DynaPro Titan operating at a wavelength of 831.2 nm. The temperature in the sample chamber could be varied between -7 and $40\,°C$ with humidity control to avoid condensation of vapor. All samples were filtered by pressing them through 0.2 μm PTFE syringe filters and were homogenized at 40 °C before measuring. The scattered light intensity was recorded for 10 s. The obtained autocorrelation curves were averaged over 10 measurements and evaluated using the cumulant expansion to the exponential decay fit model and the DynaLS-algorithm of the manufacturer. The details of the data evaluation can be found in the appendix of this chapter.

DLS was used to determine the temperature T_A, which marks the onset of agglomeration. To determine T_A the samples were cooled in 1 °C-steps from elevated temperatures. At each temperature two measurements with 10 s duration were averaged. The setup allowed cooling down to -7 °C. When the size distribution of the measured hydrodynamic radius had tripled, the samples were heated again in single degree increments with two measurements at each temperature. Heating continued until the original size distribution was reached again. The resulting hysteresis loop of the hydrodynamic radius versus the temperature was used to find the onset temperatures of agglomerate growth and decay and thus the agglomeration temperature T_A.

4.2.3 Agglomeration Kinetics

The kinetics of the agglomeration processes were monitored by quenching samples with cooling rates of ≈ 10 K/min from elevated temperatures to agglomeration temperatures below T_A in the DLS setup. At the target temperature the hydrodynamic radius was measured for 1 h or until the radius had exceeded the detection limit. An overshoot in cooling of ≈ 2 °C for ≈ 30 s was observed. The initial slope of the increase in hydrodynamic radius was used to determine the agglomeration rate constant k_{11} and the stability ratio W. The later stages of agglomeration were characterized by fitting power laws and exponential equations to the growth in hydrodynamic radius using Origin [48]. The exponent of the power laws allowed estimation of the fractal dimension of the formed agglomerates [30] (see Appendix for details.)

4.2.4 Structure Formation

After DLS measurements, 10 μl of suspension were dropped onto TEM grids and dried at ambient conditions to analyze the morphology and particle packing in the agglomerates.

The morphology of the agglomerates was also derived from the $I(q)$ profile at low q-values in SAXS measurements. The suspension was quenched to 18 and 10 °C with ≈ 30 K/min and virtually no overshoot. After resting for 1 h at the target temperature, scattered intensities from the formed sediment were measured. The detector distance was set to cover a q-range from $8.7 \cdot 10^{-2}$ to 2.3 nm^{-1}. The excess scattering at low q-values from formed agglomerates can be expressed as an additional power law term ('cluster-term') in the structure factor. From this cluster-term the fractal dimension of the formed agglomerates can be determined [46]. The details of the data evaluation can be found in the appendix of this chapter.

4.3 Results

In the following, we characterize the particles, measure agglomeration temperatures and agglomeration kinetics, and investigate the formed structures. We find identical metal cores in the suspensions, but varying agglomeration temperatures and temperature-dependent interaction potentials. These results are in agreement with a ligand-dominated interaction among the particles. The agglomeration kinetics are found to be consistent with diffusion-limited and reaction-limited mechanisms. The formed particle structures reflect the different agglomeration kinetics, as open, fractal structures are formed in diffusion-limited agglomeration and dense, globular structures in reaction-limited agglomeration.

4.3.1 Particle Characterization

The analysis of the gold concentration in the suspensions by ICP-AES resulted in an average value of 353 ± 53 mg/l. Using a gold density of 19.3 g/cm^3 [49] and the measured average radius of the gold core, a particle concentration of $(1.3 \pm 0.2) \cdot 10^{14}$ ml^{-1} can be expected. This 30 % uncertainty in particle concentration from gold concentration measurements is further increased by factors like different storage times and filtering of the samples before the different experiments. Spectroscopy measurements on the plasmon resonance of the particles (not shown) indicated reduction of particle concentrations by filtering in some instances of more than a factor of 10. Absolute particle concentrations and all values derived from it below are therefore subject to considerable uncertainties.

The particle size was determined using TEM, SAXS and DLS. Figure 4.1 gives the measured scattered intensities and autocorrelation functions of diluted samples (number density $\approx 10^{12}$ ml^{-1}) at 35 °C. The results of the fits of Eqs. 4.14 and 4.18 and the result from digital evaluation of the TEM images are given in Table 4.1. The data obtained by TEM and SAXS are in excellent agreement. The metal cores of the particles exhibit similar sizes and size distribution irrespective of the ligand used in the synthesis. DLS measurements give larger radii, a consequence of the sensitivity to the hydrodynamic radius rather than to the core size. The hydrodynamic radii increase for longer ligands and give a measure of the thickness of the ligand shell that is in good agreement with literature values for self-assembled monolayers [36]. From the data it is obvious that at least at elevated temperatures the filtered suspensions are agglomerate-free.

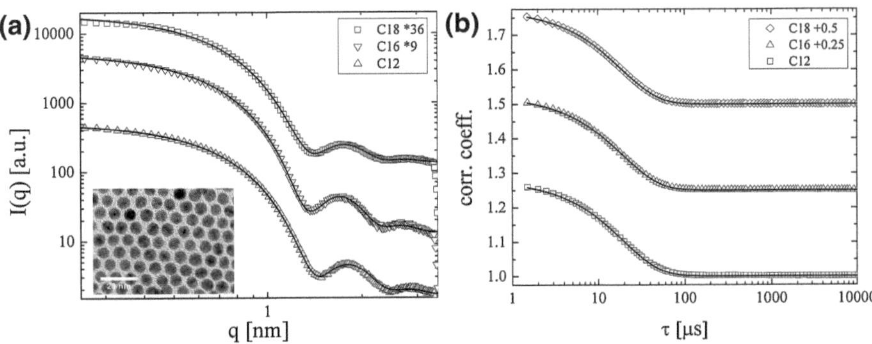

Fig. 4.1 Characterisation of the particles. **a** The scattered intensity I_q in SAXS spectra of dilute particle suspensions stabilized with C12, C16 and C18. The *solid lines* represent fits of a polydisperse particle form factor. **b** The autocorrelation functions in DLS experiments of the same suspensions. The *solid lines* represent the results of cumulant fits. The *curves* are offset by the indicated factors or addends for clarity. The inset in **a** shows a representative TEM image, the scale bar is 20 nm

Table 4.1 Results of the particle characterization by TEM, SAXS and DLS. r core radius, σ root mean square deviation (TEM and SAXS), r_H hydrodynamic radius, σ_{PD} polydispersity index (DLS)

Ligand	TEM		SAXS		DLS	
	r [nm]	σ [nm]	r [nm]	σ [nm]	r_H [nm]	σ_{PD} %
C12	3.2	0.29	3.1	0.28	4.5	3.3
C16	3.2	0.31	3.2	0.29	5.2	5.6
C18	3.2	0.32	3.1	0.28	5.3	5.4

The particles exhibit identical metal cores, the ligand shell thickness growth with ligand chain length

The IR spectra of the akyl thiol ligands in d-chloroform exhibit four peaks (insets in Fig. 4.2). These corresponds to the asymmetric and symmetric methylene stretching modes and the asymmetric in-plane and symmetric stretching modes of the terminal methyl group [50]. While the asymmetric stretching mode of the C18 ligands at $2928\,\mathrm{cm}^{-1}$ reproduces the value of liquid alkyl thiol, the value for C16 ligands is shifted to $2927.6\,\mathrm{cm}^{-1}$. This suggests that the ligands form liquid-like alkyl layers, albeit with slightly constrained motion in the case of C16 ligands. Upon cooling the d-chloroform suspensions of C18 particles and C16 particles below 30 and 24 °C, respectively, the peaks of the asymmetric stretching modes in the IR spectra shift in their position by $1\,\mathrm{cm}^{-1}$ to lower wavenumbers, indicating a higher restriction in the motion of the chains. However, the observed peak shift is minute compared to the expected shift of the peak from free alkyl thiols at $2928\,\mathrm{cm}^{-1}$ to fully frozen peak positions measured on self-assembled monolayer on planar gold surfaces at $2919\,\mathrm{cm}^{-1}$ [51]. A possible explanation for this is the high curvature of the nanoparticle surface that keeps the bundles of ligand chains from forming well-ordered packings.

From the $I(q)$ curves obtained by SAXS measurements the static structure factor $S(q)$ (Eq. 4.9), the radial pair distribution function $g(s)$ (Eq. 4.10) and the mean potential $V(s)$ (Eq. 4.11) were determined for the C16 sample. Figure 4.3a shows

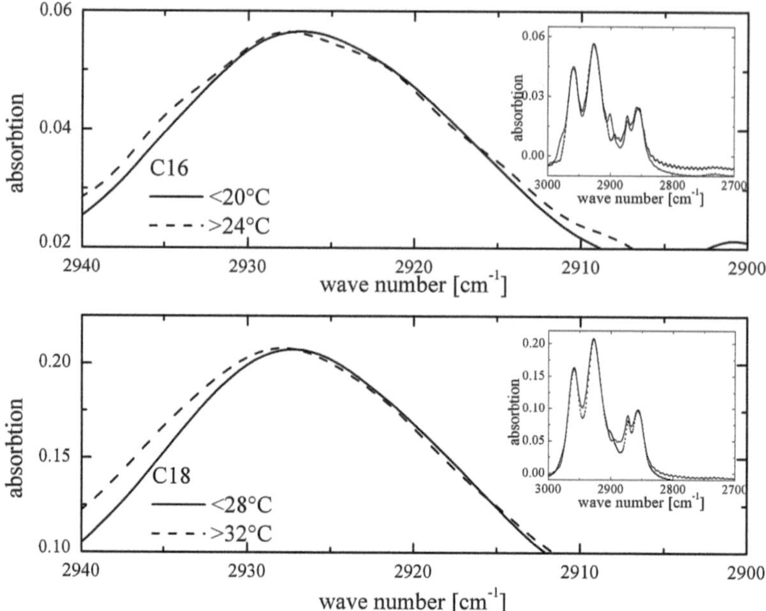

Fig. 4.2 The asymmetric methylene stretching modes measured in d-chloroform suspensions of C16 and C18 alkyl thiol gold particles at elevated and lowered temperatures. The vibration shifts towards lower wavenumber upon cooling of the suspensions. The insets give the full IR spectra, showing the characteristic vibrations of methyl and methylene groups

the evolution of the structure factor $S(q)$ when cooling from 30 to 10 °C. At the highest temperatures $S(q)$ is almost at unity, indicating an uncorrelated distribution of particles in the suspension. This is reflected by an interparticle potential with a depth of around $-1\,kT$ (Fig. 4.4). Upon cooling below 24 °C structure formation occurs in the colloidal suspension, indicated by the peak evolving at 0.74 nm^{-1}. The interparticle potential deepens below $-3/2\,kT$, the interparticle potential thus exceeds the equalizing thermal agitation. Below 16 °C remarkable higher oscillations $S(q)$ evolve, resulting in a step in $V(s)$, at which the interparticle potential deepens by 80 %. The absolute values of V_{min} have to be used with some care, as the particle concentration is a prefactor in the calculation. But the tendency indicates a strong deepening of the potential well with little cooling, proving that the particle–particle interaction is temperature-dependent.

Upon cooling the suspensions, agglomeration of the particles can be observed (Fig. 4.5). Cooling a C18 suspension below ≈ 32 °C and a C16 suspension below ≈ 24 °C induces a rapid increase in measured hydrodynamic radius. Accordingly, the hydrodynamic radius decays when heating the suspension above these temperatures, marking the respective agglomeration temperatures T_A. The C12 sample does not agglomerate in the temperatures accessible in our the setup. Storing the C12 sample in a freezer at -25 °C results in an agglomerated sample, T_A for C12 thus can be

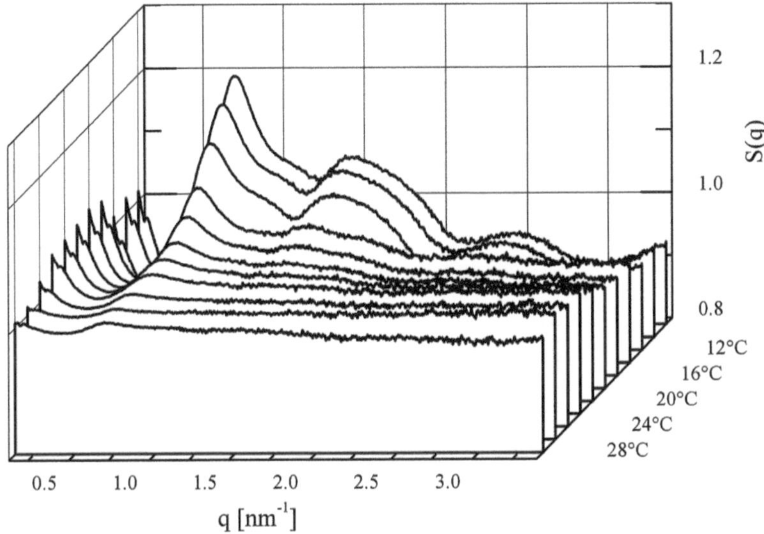

Fig. 4.3 Evolution of the structure factor $S(q)$ of a C16-suspension upon cooling. The *curves* were obtained by division of the respective I_q-curve by an I_q-curve of a dilute sample. The structure factor indicates an uniform distribution of particles at elevated temperatures. Upon cooling structures form in the suspension, leading to oscillations in the structure factor

Fig. 4.4 Trend of the depth of the effective mean inter-particle potential obtained from the structure factors in Fig. 4.3. The potential decays below $\frac{3}{2}$kT upon temperature decrease, leading to the observed structuring in the suspension. Note the distinct drop in the potential below 15 °C

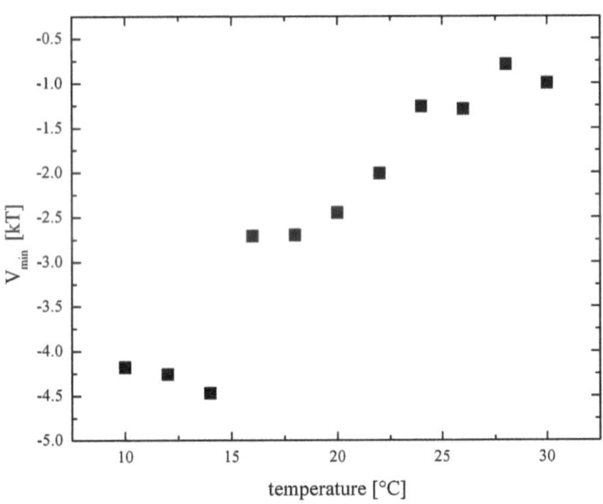

estimated between -7 and $-25\,°C$. A batch-to-batch variation of the determined T_A of a few degree Celsius have been observed. These shifts might be attributed to synthesis-to-synthesis variations in ligand coverage or to remnant solvents from synthesis or washing present in the sample, as observed in stearyl alcohol-stabilized silica particles [46].

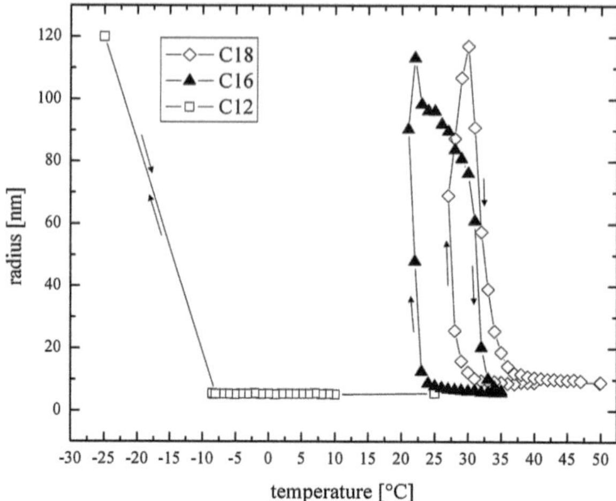

Fig. 4.5 The hydrodynamic radius measured in C12, C16 and C18 suspensions as function of temperature. The *arrows* indicate the heating/cooling cycle. The increase of the hydrodynamic radius with cooling indicates the onset of agglomeration, the decay with heating the disassembling of agglomerates. The temperature of agglomeration onset grows markedly with the ligand chain length

4.3.2 Agglomeration Kinetics

The initial agglomeration kinetics of C16 samples quenched to temperatures below T_A as measured by DLS are displayed in Fig. 4.6a. Above 14 °C the agglomeration rate increases with decreasing agglomeration temperature, below 14 °C the increase in agglomeration rate stagnates. Calculating the stability ratio (Fig. 4.6b) using Eqs. 4.22 and 4.25, the transition from temperature-dependent agglomeration to nearly temperature-independent agglomeration becomes apparent. The absolute value of the largest determined rate constant k_{11} calculates to $2.9 \cdot 10^7$ nm/s using the measured particle concentration. This is three orders of magnitude lower than the theoretical limit of diffusion-limited agglomeration $k_{11}^{Smol} = 2.7 \cdot 10^{10}$ nm/s (Eq. 4.24). This might be an effect of the shallower attractive well in the sterically stabilized suspension compared to the infinitely deep potential well assumed for SMOLUCHOWSKIAN agglomeration. The uncertainty in the particle concentration used for calculating the rate constants might contribute as well to the deviation. Despite the uncertainties, two different regimes in agglomeration kinetics are clearly resolved.

The intermediate growth kinetics are characterized by the transition from particle-particle to agglomerate-agglomerate aggregation. The measurements supports a difference in growth mechanism between the temperature-dependent agglomeration above and the temperature-independent agglomeration below 15 °C. At 14 °C the

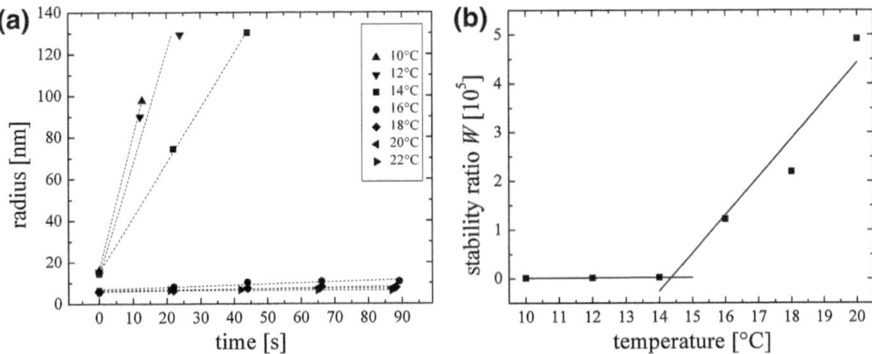

Fig. 4.6 The initial agglomeration kinetics of C16 samples below T_A. **a** The first hydrodynamic radii measured after quenching the sample to target temperatures. The *lines* are linear least *square* fits to each temperature's data points. The growth rate increases with decreasing temperatures. **b** The stability ratio calculated from the growth rates and the theoretical SMOLUCHOWSKY rate constant. The *lines* are linear least *square* fits to the leftmost three data points and the rightmost four data points, respectively. Below 15 °C the particle agglomeration becomes much less temperature sensitive. Even the highest experimental rate constants is at 1/1000 of the theoretical maximum rate constant (see text for values)

Fig. 4.7 The intermediate growth kinetics of C16 samples. The difference in agglomeration kinetics can be seen from the agglomerate growth: at 16°C the agglomerates grow exponentially (*solid line*), at 14 °C power-law-like (*solid line*). The inset is a double logarithmic plot of the data at 14 °C that illustrates the power law behavior

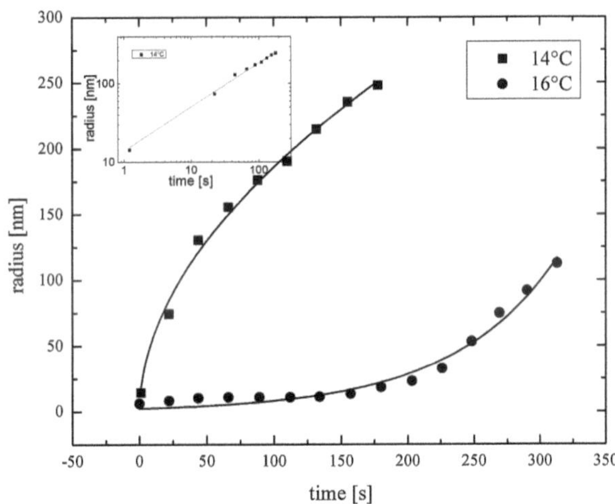

increase in hydrodynamic radius follows a power law, while at 16 °C the increase is exponential (Fig. 4.7). The exponential growth is characteristic for reaction-limited growth, whereas power-law kinetics indicate diffusion-limited growth (Eqs. 4.27 and 4.26). An exponent of 0.56 can be obtained from the power law fit to the data at 14 °C using Eq. 4.26. This indicates a fractal dimension of 1.79 of the formed agglomerates.

The exponential growth of the agglomerates at temperatures above 15 °C finally changes to power-law like growth (Fig. 4.8). This indicates the consumption of primary particles so that the agglomeration becomes limited by the slow diffusion of

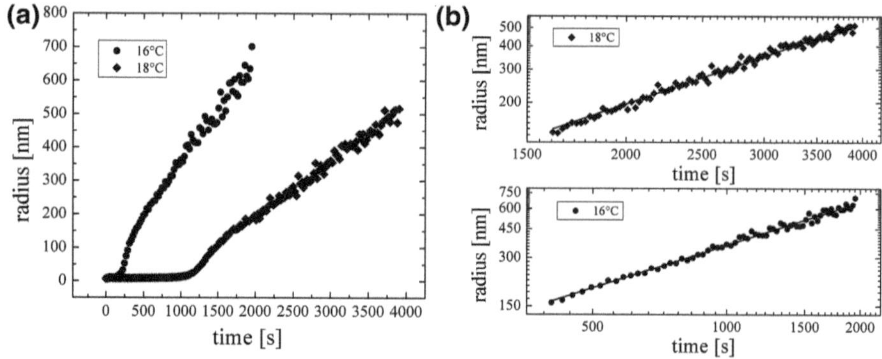

Fig. 4.8 The final growth kinetics of C16 samples above 15 °C. **a** Evolution of the agglomerate growth. **b** Double logarithmic plot of the final growth regime, emphasizing that the growth approximately follows a power law. The *solid lines* are least-square fit of power laws

large agglomerates. A very low fractal dimension of the agglomerates of ≈ 1.25 can be derived from the power-law fit.

4.3.3 Structure Formation

In Fig. 4.9 the scattered intensities from a dilute suspension are compared to the intensity profiles of samples that were kept for one hour at 10 and 18 °C. A change of the profiles at low temperatures for small q-values is observable. The fit of the profiles using a particle form factor $F(q)$ (Eq. 4.14) and a power-law cluster-term $S_c(q)$ (Eq. 4.15) yields the exponents and the fractal dimension of the formed agglomerates. At 18 °C the profile is best fit with an exponent of 1.3, indicating dense structures with smooth surfaces and a fractal dimension of 2.6. In contrast, at 10 °C the best fit has an exponent of 0.788, thus the cluster have much more rough surface, represented by a low fractal dimension of 1.58. Peaks in the intensity profile arising from BRAGG diffraction from crystalline packings are not observable.

TEM micrographs underline the a markedly different morphology of the agglomerates at the respective temperatures (Fig. 4.10). At temperatures below 15 °C, the agglomerates have an open filamentous or network-like structure. At temperatures above 15 °C, the agglomerates are rather dense and globular. At later stages of the growth process, the agglomerates join to form more open superstructures again. The packing of the particles inside the agglomerates is dense and irregular. Ordered packings were not observable at any of the investigated temperatures.

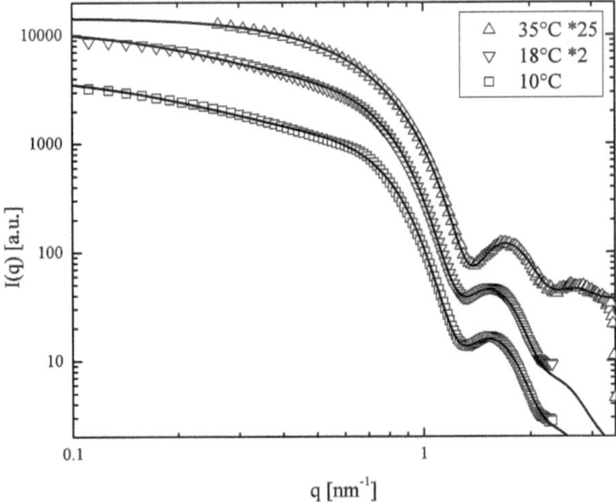

Fig. 4.9 Scattered intensities I_q from SAXS measurements of the dilute suspension of C16 particles at 35 °C, and suspensions that rested for 1 h at the denoted temperatures. The samples at lower temperatures exhibit altered profiles due to scattering from agglomerates. The *solid lines* represent fits to the profiles using the form factor alone for the dilute sample and the form factor and a cluster term for the samples at the lower temperatures. The *curves* are offset by the indicated factors for clarity

4.4 Discussion

The particle characterization with TEM and SAXS prove identical metal cores for the three syntheses with C12, C16 and C18 ligands. The contribution to the particle interaction originating from the cores is virtually identical for all samples.

The stability of the particle suspensions, in contrast, strongly deviates. The agglomeration temperatures of the samples depend on the chain length; the suspension with the longest ligands is destabilized at the highest temperature. The temperature at which a C12 sample agglomerates is ≈ 45 °C lower than the agglomeration temperature of a C18 sample. The longer the ligands are, the stronger the attraction among the particles appears to be. Additionally, the interaction potential between the particles is temperature-dependent. The depth of the interparticle potential measured with SAXS in a C16 sample growths by ≈ 350 % from 24 °C to 14 °C. It is reasonable to assume the gold cores to be unchanged in this temperature regime. The shift in the interaction must therefore be a result of a change in the ligand shell.

From this results we conclude that the ligand-ligand interactions dominate and solubility is the main mechanism in particle assembly. The alkyl thiol ligand shells form a thin layer of highly concentrated polymer solution around the particles. This polymer solution then follows the predictions of the FLORY- HUGGINS-theory [52]. Entropic effects prevail in the thin polymer shell above the θ-temperature of the respective polymer-solvent pair, the chains effectively repel each other and stabilize the suspension. Cooling below the θ-temperature, the polymer chains attract each

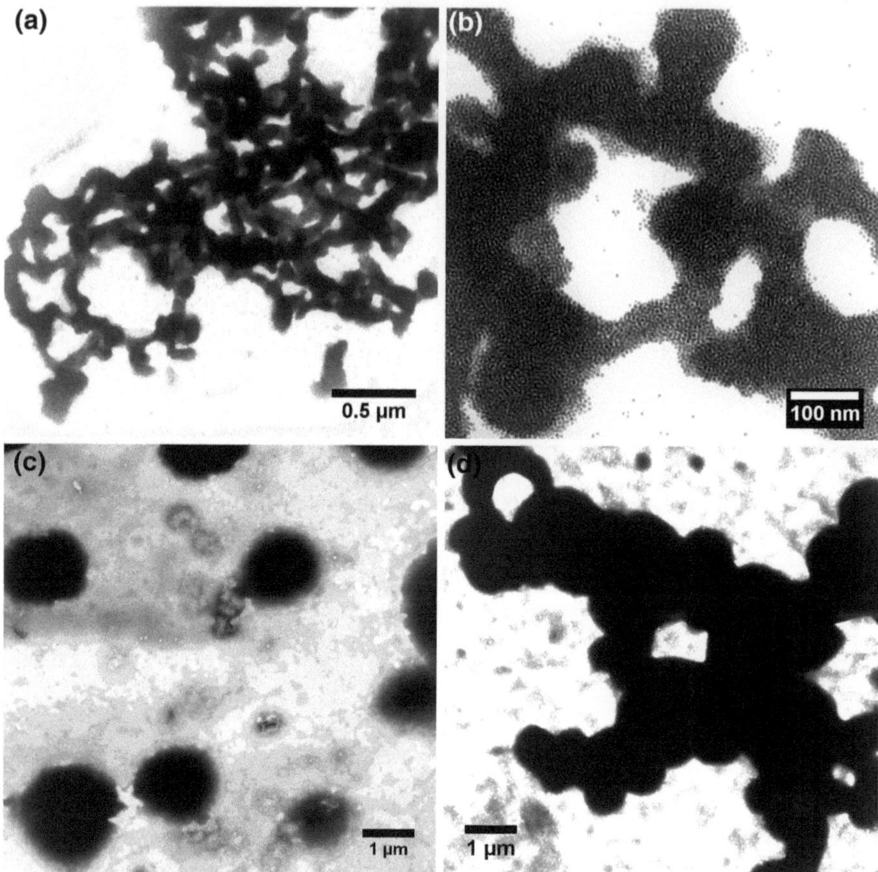

Fig. 4.10 TEM micrographs of agglomerates of C16 particles. **a** and **b** Agglomerates formed at 10 °C after 5 min growth, exhibiting filamentous structures after diffusion-limited growth, **c** agglomerates formed at 16 °C after 5 min growth, showing the formation of globular structures after reaction-limited growth, and **d** agglomerates at 16 °C after 30 min growth, showing coalescence of agglomerates

other and the particles become attractive. In the example of the C16 particles a θ-temperature of slightly above 22 °C can be expected, and the attraction suffices to cause agglomeration at 22 °C.

The jump in the interaction potential at 15 °C, however, is not consistent with a smooth variation in the polymer chain solubility with temperature. A transition in the particle ligand shells seems likely. Such a transition has been measured for stearyl alcohol stabilized silica particle suspensions [34, 45]. The ligands change their conformation upon cooling from a random, solvated orientation to densely packed, straight chains. This increase in packing density is assumed to be accompanied by an increase of the VAN DER WAALS-attraction between the shells [34, 45]. We assume

Table 4.2 Characteristic temperatures: melting temperatures T_m of bulk alkane and bulk alkyl thiol [49], melting temperatures T_m of alkyl thiols anchored on 2 nm gold nanoparticles [53], and the measured agglomeration temperatures T_A of alkyl thiol-stabilized suspension

Chain length	T_m alkane	T_m alkyl thiol	T_m alkyl thiol ligands	T_A gold nanoparticles
C12	−10	−7	≈ 3	Between −25 and −7
C16	18	19	≈ 41	≈ 22
C18	28	30	≈ 51	≈ 32

All temperatures are given in °C

therefore that the temperature evolution of the particle interaction can be described by two temperatures, the θ-temperature of the polymer-solvent pair, below which the particles turn attractive, and the phase-transition temperature of the ligand shell, below which the ligand shell undergo a packing transition and increase the attraction.

Both temperatures depend on the magnitude of the interactions among the ligand chains. A good indicator for the interactions among the chains is their bulk melting temperature. Table 4.2 gives the melting temperatures of bulk alkanes and of bulk alkyl thiols, the melting temperatures of the respective alkyl thiol ligand shell on gold nanoparticles and the measured agglomeration temperatures. The melting temperatures of the alkyl thiol ligand shells measured in air are increased compared to their bulk alkyl thiol counterparts due to the immobilization on the gold surfaces. The agglomeration temperatures of the particles in suspension nevertheless coincide well with the melting temperatures of the bulk counterpart of the ligands. The melting point of the bulk compound can thus be used to characterize the stability of alkyl thiol-stabilized suspensions. An interesting question is how far this behavior can be extrapolated to shorter chain lengths. A particle suspension stabilized with heptyl thiol should be stable down to unrealistic −91 °C, where the solvent freezes.

For the sake of completeness it has to be added that the measured agglomeration temperature of the octadecyl thiol (C18) stabilized sample in heptane is ≈ 20 °C higher than the reported agglomeration temperatures of stearyl alcohol stabilized silica particles suspended in hexane or heptane which agglomerate below 10 °C [41, 44]. This might be due to the attractive contribution of the gold cores, although the even lower agglomeration temperature of C12 gold nanoparticles render this explanation unlikely. More likely this is an effect of the different surface coverage with ligands. Self-assembled monolayer (SAM) on silica are known to form less dense packings than alkyl thiol-SAM on gold [54]. The ligand shells of silica particles in suspension therefore resemble less dense polymer solutions with a lower free energy of mixing compared to the thiol counterpart. Additional charge stabilization from unsaturated surface groups on the silica particles are also discussed [55].

The kinetics of the agglomerate growth reflect the evolution of the interparticle potential and the particle solubility with temperature. When cooling the suspension below the θ-temperature of the solvent-ligand pair, the particles start to agglomerate with reaction-limited characteristics. Further cooling lowers the solubility of the ligands, causing an increase of the agglomeration rate and a decrease in the stability ratio. The transition in the ligand shell deepens the attractive potential such that every encounter of particles leads to aggregation, which turns diffusion-limited. Importantly, this transitions from a stable suspension via reaction-limited agglomeration to diffusion-limited agglomeration depends only on the solubility and packing of the ligand chains on the individual particles. This is not fully equivalent to phase transitions in systems with temperature-independent interactions like VAN DER WAALS-interactions. In such systems phase transition by bimodal or spinodal mechanisms can be observed, which might lead to similar agglomeration kinetics. However, the binodal and spinodal transitions are highly particle concentration depended. In the suspensions considered in this work the stability is independent of the particle concentration, but dependent on the concentration in the polymer solution formed by the ligand shell of each particle.

The increase of the stability ratio with temperature, opposite to the behavior expected for charge-stabilized suspension, indicates the absence of kinetic repulsive barriers. The continued growth of the agglomerates, however, follows the rules developed for charge-stabilized suspensions. In the diffusion-limited regime the agglomerates grow power-law like, in the reaction-limited regime they grow exponentially. Eventually, the slow diffusion of large agglomerates turns the reaction-limited agglomeration diffusion-limited. From the agglomeration kinetics measured with DLS already some information on structure formation processes can be gained. From the exponent of the power-law fit to the diffusion-limited growth a fractal dimension of 1.79 can be estimated. The low fractal dimension indicate the formation of open agglomerate structures with rough surfaces. The value of 1.79 is similar to values measured on charge-stabilized gold nanoparticle suspensions [28, 31] but smaller than the values reported for sterically stabilized silica [44, 46].

The fit to the SAXS scattering curves of agglomerated suspensions supports the formation of two different structures in the two different agglomeration kinetics. A fractal dimension of 1.58 is calculated for the agglomerates grown at 10 °C, in contrast to the smoother and denser agglomerates formed at 18 °C, as indicated by the a fractal dimension of 2.6. TEM images finally show that different agglomerate structures are preserved even during sample preparation, which involves pipetting and drying of the suspension. The difference in the agglomerate morphologies formed in the two different growth kinetics is remarkably. Below 15 °C the agglomerates are interconnected filamentous structures, similar to structures formed by spinodal decomposition [56]. Above 15 °C the formed agglomerates are globular, suggesting a nucleation-and-growth mechanism. The formed globular agglomerates coalesce with continuing agglomeration and form interconnected structures. The typical length scale of these structures is ten times larger than the sizes of the structures grown with diffusion-limited kinetics. The agglomeration kinetics can thus be used to direct the structure formation in agglomeration of sterically stabilized suspension.

Puzzling is the absence of sharp BRAGG-peaks in the structure factors measured with SAXS, which are characteristic for colloidal crystals [50, 57]. In TEM, no indication for colloidal crystallization was found in the micrographs, too. At this point we can only speculate over the absence of crystallization in monodispere particle suspensions, which have already been reported to form colloidal crystals upon precipitation or evaporation [11].

The agglomeration of the particles marks a phase separation between a stable particle suspension and a condensed particle phase. This phase separation might be between a stable 'gas-like' low particle density fluid phase and a 'liquid-like' high particle density fluid phase, rather than between a fluid and a solid phase. The drying of the agglomerated suspension would then vitrify the liquid-like phase, and the TEM shows snapshots of the unordered liquid agglomerates. However, simulations have already shown that in thermodynamic phase diagrams of suspensions with short-ranged interactions the fluid-fluid phase separation line is buried under the fluid-solid phase separation line and a glass transition line [58–60]. The attractions present below the θ-temperature stem from a lowering of the solubility of the ligand chains. The particles thus lower their surface energy upon contact below the θ-temperature, what causes a short-ranged attraction, which is limited to the length of the ligands. The phase separation thus can be assumed to be between a fluid and a solid phase of the particle suspension. More likely, the absence of order is an effect of the overlap in ligand shell upon contact. Due to the attraction of the ligands the ligand shells will interdigitate upon contact. The entangled ligand chains form an interlock against fast rearrangement, and the solidification of the condensed phase preserve the stochastic packing of the particles. The mechanisms preventing crystallization and differences between the experiments presented here and the precipitation experiments leading to crystallinity will be evaluated in Chap. 5.

4.5 Conclusion

This work reveals three important features of alkyl thiol-stabilized metal particle suspensions. First, the particle characterization shows that the contribution of the ligand chains to the interparticle potential prevails over the contribution of the metal core at least in the range of ligand chain lengths investigated. The stability of the suspension can be deduced from the melting temperature of the bulk compound of the ligands, whereas a low melting point provides a good stability. Second, the kinetics of the agglomeration follow the laws developed for diffusion-limited and reaction-limited agglomeration in charge-stabilized suspensions. The agglomeration kinetics define the morphology of the formed agglomerates. In diffusion-limited agglomeration filamentous networks of particles are formed, in reaction-limited agglomeration globular structures. Third, by inducing agglomeration by cooling alkyl thiol-stabilized metal nanoparticle suspensions no colloidal crystals are formed, irrespective of the growth kinetics.

4.6 Appendix

4.6.1 Small-Angle X-ray Scattering Data Evaluation

The scattered intensity I_s measured in a small-angle scattering experiment at each detector element defining an solid angle $\Delta\Omega$ is given by [61]

$$I_s = I_0 \cdot \epsilon \cdot T_t \cdot \Delta\Omega \cdot A_s \cdot l_s \cdot \frac{d\Sigma}{d\Omega}, \tag{4.6}$$

where I_0 is the incident photon intensity per unit area per unit time, ϵ the detector efficiency, $T_t = \frac{I_t}{I_0}$ the sample transmission with I_t the transmitted intensity per unit area per unit time, A_s the cross section of the beam and l_s is the sample thickness. The quantity $\frac{d\Sigma}{d\Omega}$ is the differential scattering cross section per unit volume and contains the information about the structure of the sample. The quantitative evaluation of the measured intensities therefore requires a normalization of the data to $\frac{d\Sigma}{d\Omega}$. But $\epsilon \cdot T_t \cdot l_s$ is a quantity unknown a priori in our experiments which limited evaluation possibilities. To still gain information, we kept parameters influencing $\epsilon \cdot T_t \cdot l_s$ constant during the experiments, leaving the incident intensity I_0 and $\frac{d\Sigma}{d\Omega}$ the only variables, and only used the information contained in the relative intensity:

$$I_r = \frac{\tilde{I}_{s,1}}{\tilde{I}_{s,2}} = \frac{\left(\frac{d\Sigma}{d\Omega}\right)_1}{\left(\frac{d\Sigma}{d\Omega}\right)_2}, \tag{4.7}$$

where $\tilde{I}_s = \frac{I_s}{I_0}$ is the scattered intensity normalized by the incident photon intensity.

The differential cross section of a suspension of spherical particles with narrow size distribution can be expressed as a product of a form factor $F(q)$ and a structure factor $S(q)$ using a scattering vector q [62]:

$$\frac{d\Sigma}{d\Omega} = \varrho \cdot F(q) \cdot S(q), \tag{4.8}$$

where the factor ϱ accounts for the particle volume fraction and the scattering contrast between the particles and the solvent. The norm of the scattering vector q is given by $4\pi/\lambda \cdot \sin(\Theta/2)$ with λ the wavelength of the scattered light and Θ the scattering angle. $F(q)$ describes the scattering by individual particles and contains information about size and shape of the particles. $S(q)$ describes the interference by light scattered from different particles and contains information about the interparticle interactions. In dilute suspensions $S(q) \approx 1$. By calculating the relative intensity using the scattered intensity from a dilute sample as denominator, the structure factor can be extracted:

$$I_r = \frac{\tilde{I}_s}{\tilde{I}_{s,dilute}} = \frac{\varrho \cdot F(q) \cdot S(q)}{\varrho_{dilute} \cdot F(q)} = \frac{\varrho}{\varrho_{dilute}} \cdot S(q). \tag{4.9}$$

The pair correlation function $g(s)$, the probability of finding a particle at a distance s from another particle, can be calculated from $S(q)$ by a FOURIER-transform [63]:

$$g(s) = 1 + \frac{1}{2\pi s \rho} \int_0^\infty q(S(q) - 1) \sin(qs) dq, \qquad (4.10)$$

where ρ is the particle number density. With g(s) it is possible to obtain information about the interparticle interaction [42]:

$$g(s) = e^{V(s)/kT}, \qquad (4.11)$$

where $V(s)$ is the so-called "potential of mean force", the potential of two particles in the field of all other particles.

An alternative approach to extract information from I_s is to fit analytical expressions to the intensity profile. Merging the experimental factors in Eq. 4.6 and ϱ in Eq. 4.8 to a single prefactor or fit parameter yields:

$$I_s = \xi \cdot F(q) \cdot S(q). \qquad (4.12)$$

Again, for dilute samples $S(q)$ becomes ≈ 1, and the profile of the scattered intensity $I(q)$ becomes solely a function of the particle size and shape. Assuming a profile for the form factor and fitting it to the measured intensity yields information on the particle size and size distribution. For homogeneous spherical particles a mathematical expression for the form factor is long known [64]:

$$F(qr) = \left[3 \cdot \frac{\sin(qr - qr) \cdot \cos(qr)}{(qr)^3} \right]^2, \qquad (4.13)$$

where r is the radius of the particles. For polydisperse particles, the size distribution $(n(r)/n_{total})$ has to be considered [65]:

$$I_s = \xi \cdot \int_0^\infty \left(\frac{n(r)}{n_{total}} \right) F(qr) r^6 dr \qquad (4.14)$$

An analytical expression for Eq. 4.14 is available for a SCHULTZ-distribution of the particle sizes [62].

When the particles become sufficiently attractive, they form clusters which give rise to excess scattering at low q. The excess scattering can be considered in the fitting procedure by adding a power-law term [46]:

$$S_c(q) = \frac{a}{(1 + b \cdot q^2)^p}, \qquad (4.15)$$

where a and b are fit parameters linked to the mass and size of the clusters. The exponent p of the power-law is related to the fractal dimension of the clusters d_f:

$$p \approx d_f/2. \tag{4.16}$$

The actual fitting of the size and size distribution of the particle form factor and the exponent of the cluster term was performed using the SAXS-utilities software developed by M. SZTUCKI at ESRF [66].

4.6.2 Dynamic Light Scattering and Agglomeration Kinetics

The temporal fluctuation of the intensity of light scattered by a colloidal suspension from a coherent laser beam measured by a photo multiplier can be characterized by the intensity autocorrelation function

$$C(q, \tau) = \langle n(q, t)n(q, t + \tau) \rangle, \tag{4.17}$$

where q is the scattering vector, n is the number of photon counts in a time interval between t and $t + \delta t$ and τ a delay time. The brackets indicate averaging over the total sampled time. Following the derivation given by BERNE and PERCORA [67] the autocorrelation function for GAUSS-distributed scattered light intensities takes the form

$$C(q, \tau) = \langle n \rangle^2 \left[1 + f \cdot e^{\frac{-2\tau}{\tau_D}} \right] \equiv \langle n \rangle^2 \left[1 + f \cdot g_1^2 \right], \tag{4.18}$$

where $\langle n \rangle^2$ is the average number of counts in the time interval δt, f a spatial coherence factor which depends on the detector area and the sampling interval δt. τ_D is a relaxation time connected to the diffusion coefficient D of the colloidal particles:

$$\tau_D = \frac{1}{q^2 \cdot D}. \tag{4.19}$$

From the diffusion coefficient a representative particle radius r_{mean} can be determined using the STOKES- EINSTEIN relation:

$$D = \frac{kT}{6\pi \eta r_{mean}}, \tag{4.20}$$

where η is the solvent viscosity. As real colloids always exhibit a size distribution, the radius r_{mean} determined from Eqs. 4.19 and 4.20 gives an average radius of all species present in the scattering volume. To gain information on the size distribution in the sample, a cumulant analysis can be applied [68]. Expanding the exponential function g_1 in a power series in τ yields

$$g_1 = e^{(\frac{-\tau}{\tau_D})(1+\frac{k_2}{2!}\cdot\tau^2-\frac{k_3}{3!}\cdot\tau^3+...)}, \qquad (4.21)$$

where τ_D gives the average diffusion coefficient and $k_2 \cdot \tau_D^2$ the polydispersity index of the suspension. For a GAUSSIAN distribution of colloid sizes all higher cumulants are equal to zero.

The agglomeration rate constant k_{11} for the formation of doublets from initial singlets (free particles) from SMOLUCHOWSKI'S theory of particle agglomeration [26] is defined in Eq. 4.1. VIRDEN and BERG[69] have developed the relation between the growth of the mean radius measured with dynamic light scattering r_{mean} and the agglomeration rate constant k_{11}:

$$- k_{11} = \frac{1}{r_{mean,0}n_0\alpha}\frac{dr_{mean}(t)}{dt}, \qquad (4.22)$$

where $r_{mean,0}$ is the initial mean radius of the particles, n_0 the initial particle number concentration and α is an optical factor, which the authors derive from RAYLEIGHT-DEBYE scattering theory. A possibility to determine k_{11} without the use of optical factors requires the combination of static and dynamic light scattering [70, 71]:

$$\frac{1}{I(0)}\frac{dI(t)}{dt} = \left(1 - \frac{r_1}{r_2}\right)^{-1}\frac{1}{r_{mean,0}}\frac{dr_{mean}(t)}{dt} - k_{11}\cdot n_0, \qquad (4.23)$$

where $I(0)$ is the initial scattered intensity and r_1/r_2 is the ratio between the hydrodynamic radius of a singlet and a doublet. Both methods require a precise knowledge of the initial particle number concentration n_0, which can be cumbersome for real colloidal suspensions.

The determined rate constant can be compared to the rate constant of purely diffusion controlled, fast, or SMOLUCHOWSKIAN agglomeration:

$$k_{11}^{Smol} = 16\pi r D = \frac{8kT}{3\eta} \qquad (4.24)$$

The ratio W between the experimental rate constant and the fast rate constant is called the stability ratio of the suspension, and contains information on the particle interaction [27].

$$W = \frac{k_{11}^{Smol}}{k_{11}} \qquad (4.25)$$

For charge-stabilized suspensions, where the agglomeration rate is solely determined by the rate of particles hopping over an energy barrier of height V_{max} into an infinite deep potential well, so no back-reaction of desorbing particles, the stability ratio can be written as $W \approx e^{\frac{V_{max}}{kT}}$.

WEITZ and LIN et al. have investigated in a series of publication the relation between the mean radius r_{mean} measured by DLS and the fractal dimension of formed agglomerates and with the growth kinetics in the suspension [22, 28–31, 72–74]. For diffusion-limited or fast agglomeration, they found a power-law like growth of the mean radius with time

$$r_{mean} \propto t^{\alpha},\tag{4.26}$$

where α is the inverse fractal dimension $d_f = 1/\alpha$. The power-law shape with an exponent smaller 1 can be understood qualitatively as diffusion is the limiting parameter for fast coagulation. Every collision leads to sticking, and reduces the diffusion coefficient of the formed object. The decreasing diffusion slows down the reaction. The relation to the fractal dimension can be derived from the definition of fractality. The mass M of an agglomerate is given by the radius of gyration, the radius of a single particle and the fractal dimension: $M \propto (R_g/r_1)^{d_f}$. For a solid object d_f would equal 3. From SMOLUCHOWSKI'S theories a linear growth of the average agglomerate mass with time is predicted, $\overline{M} = 1 + t/t_0$, whereas t_0 is determined by the rate constant k_{11}^{Smol} and the initial particle concentration. With $r_{mean} \propto R_g$ above relation follows.

For reaction-limited or slow agglomeration, they found an exponential dependency of the mean radius with time

$$r_{mean} \propto e^{t/t_0},\tag{4.27}$$

where t_0 is a sample dependent time constant. The exponential behavior can be understood in a qualitative sense as the sticking probability upon contact of particles limits slow agglomeration. Larger objects have a higher number of touching points upon collision. The sticking probability will therefore increase with object size and increase the reaction rate. This exponential growth is ultimately limited by the slowing down of the diffusion with growing object size, rendering the agglomeration diffusion-limited for long agglomeration times.

References

1. M. Antonietti, C. Göltner, Superstructures of functional colloids: chemistry on the nanometer scale. Angew. Chem. Int. Ed. Engl. **36**, 910–928 (1997)
2. M.-C. Daniel, D. Astruc, Gold nanoparticles: assembly, supramolecular chemistry, quantum-size-related properties, and applications toward biology, catalysis, and nanotechnology. Chem. Rev. **104**, 293–346 (2004)
3. T.K. Saun, A.L. Rogach, F. Jäckel, T.A. Klar, J. Feldmann, Properties and applications of colloidal nonspherical noble metal nanoparticles. Adv. Mater. **22**, 1805–1825 (2010)
4. U. Gasser, Crystallization in three- and two-dimensional colloidal suspensions. J. Phys. Conden. Matter **21**, 203101 (2009)
5. B.L.V. Prasad, C.M. Sorensen, K.J. Klabunde, Gold nanoparticle superlattices. Chem. Soc. Rev. **37**, 1871–1883 (2008)
6. S. Kinge, M. Cego-Calama, D.N. Reinhoudt, Self-assembling nanoparticles at surfaces and interfaces. ChemPhysChem **9**, 20–42 (2008)

7. F. Li, D.P. Josephson, A. Stein, Kolloidale organisation: der Weg vom Partikel zu kolloidalen Molekülen und Kristallen. Angew. Chem. **123**, 378–409 (2011)
8. A. Yethiraj, Tunable colloids: control of colloidal phase transitions with tunable interactions. Soft Matter **3**, 1099–1115 (2007)
9. C.B. Murray, C.R. Kagan, M.G. Bawendi, Synthesis and characterization of monodisperse nanocrystals and close-packed nanocrystal assemblies. Ann. Rev. Mater. Sci. **30**, 545–610 (2000)
10. D.V. Talapin, J.-S. Lee, M.V. Kovalenko, E. Shevchenko, Prospects of colloidal nanocrystals for electronic and optoelectronic applications. Chem. Rev. **110**, 389–458 (2010)
11. N. Zheng, J. Fan, G.D. Stucky, One-step one-phase synthesis of monodisperse noble-metallic nanoparticles and their colloidal crystals. J. Am. Chem. Soc. **128**, 6550–6551 (2006)
12. D.V. Talapin, E.V. Shevchenko, A. Kornowski, N. Gaponik, M. Haase, A.L. Rogach, H. Weller, A new approach to crystallization of CdSe nanoparticles into ordered three-dimensional superlattices. Adv. Mater. **13**, 1868–1871 (2001)
13. D.V. Talapin, E.V. Shevchenko, C.B. Murray, A. Kornowski, S. Förster, H. Weller, CdSe and CdSe/CdS nanorod solids. J. Am. Chem. Soc. **126**, 12984–12988 (2004)
14. C.B. Murray, C.R. Kagan, M.G. Bawendi, Self-organization of CdSe nanocrystallites into three-dimensional quantum dot superlattices. Science **270**, 1335–1338 (1995)
15. A. Dong, X. Ye, J. Chen, C.B. Murray, Two-dimensional binary and ternary nanocrystal superlattices: the case of monolayers and bilayers. Nano Lett. **11**, 1804–1809 (2011)
16. B.A. Korgel, D. Fitzmaurice, Condensation of ordered nanocrystal thin films. Phys. Rev. Lett. **80**, 3531–3534 (1998)
17. M.B. Sigman, A.E. Saunders, B.A. Korgel, Metal nanocrystal superlattice nucleation and growth. Langmuir **20**, 978–983 (2004)
18. T.P. Bigioni, X.-M. Lin, T.T. Nguyen, E.I. Corwin, T.A. Witten, H.M. Jaeger, Kinetically driven self assembly of highly ordered nanoparticlemonolayers. Nat. Mater. **5**, 265–270 (2006)
19. C. Lu, Z. Chen, S. O'Brien, Optimized conditions for the self-organization of CdSe-Au and CdSe-CdSe binary nanoparticle superlattices. Chem. Mate. **20**, 3594–3600 (2008)
20. J.S. Langer, Instabilities and pattern formation in crystal growth. Rev. Mod. Phys. **52**, 1–28 (1980)
21. G. Tegze, G.I. Toth, L. Granasy. Faceting and branching in 2D crystal growth. Phys. Rev. Lett. **106**, 195502-1–195502-4 (2011)
22. M.Y. Lin, H.M. Lindsay, D.A. Weitz, R.C. Ball, R. Klein, P. Meakin, Universality in colloidal aggregation. Nature **339**, 360–362 (1989)
23. A.T. Skjeltorp, Visualization and characterization of colloidal growth from ramified to faceted structures. Phys. Rev. Lett. **58**, 1444–1447 (1987)
24. W.B. Russel, D.A. Saville, W.R. Schowalter, *Colloidal Dispersions* (Cambridge University Press, Cambridge, 1989)
25. J.J. Cerda, T. Sintes, C.M. Sorensen, A. Chakrabarti, Kinetics of phase transformations in depletion-driven colloids. Phys. Rev. E **70**, 011405-1–011405-9 (2004)
26. M.V. Smoluchowski, Versuch einer mathematischen Theorie der Koagulationskinetik kolloider Lösungen. Zeitschrift für physikalische Chemie **92**, 129–168 (1917)
27. D. Myers, *Surfaces, Interfaces, and Colloids: Principles and Applications* (Wiley, New York, 1999)
28. M.Y. Lin, H.M. Lindsay, D.A. Weitz, R. Klein, R.C. Ball, P. Meakin, Universal diffusion-limited colloidal aggregation. J. Phys. Conden. Matter **2**, 3093–3113 (1990)
29. D.A. Weitz, J.S. Huang, M.Y. Lin, J. Sung, Dynamics of diffusion-limited kinetic aggregation. Phys. Rev. Lett. **53**, 1657–1660 (1984)
30. D.A. Weitz, M.Y. Lin, C.J. Sandroff, Colloidal aggregation revisited: new insights based on fractal structure and surface-enhanced raman scattering. Surf. Sci. **158**, 147–164 (1985)
31. M.Y. Lin, H.M. Lindsay, D.A. Weitz, R.C. Ball, R. Klein, P. Meakin, Universal reaction-limited colloidal aggragation. Phys. Rev. A **41**, 2005–2020 (1990)
32. W.D. Luedtke, U. Landman, Structure, dynamics, and thermodynamics of passivated gold nanocrystallites and their assemblies. J. Phys. Chem. **100**, 13323–13329 (1996)

33. C. Gutierrez-Wing, P. Santiago, J. Ascencio, A. Camacho, M. Jose-Yacaman, Self-assembling of gold nanoparticles in one, two, and three dimensions. Appl. Phys. A **71**, 237–243 (2000)
34. S. Roke, O. Berg, J. Buitenhuis, A. van Blaaderen, M. Bonn, Surface molecular view of colloidal gelation. Proc. Nat. Acad. Sci. **103**, 13310–13314 (2006)
35. A.K. Doland, S.F. Edwards, Theory of the stabilization of colloids by adsorbed polymer. Proc. R. Soc. Lond. A Math. Phys. Sci. **337**, 509–516 (1974)
36. A. Ulman, An introduction to ultrathin organic films: from Langmuir-Blodgett to self-assembly (Academic Press, New York, 1991)
37. V.A. Parsegian, Van der Waals forces (Cambridge University Press, Cambridge, 2006)
38. J. Lyklema, *Fundamentals of Interface and Colloid Science. Volume I: Fundamentals* (Academic Press, New York, 1991)
39. J.W. Jansen, C.G. Kruif, A. Vrij, Phase seperation of sterically stabilized colloids as a function of temperature. Chem. Phys. Lett. **107**, 450–453 (1984)
40. J.W. Jansen, C.G. Kruif, A. Vrij, Attractions in sterically stabilized silica dispersions I. Theory of phase separation. J. Colloid Interface Sci. **114**, 471–480 (1986)
41. J.W. Jansen, C.G. Kruif, A. Vrij, Attractions in sterically stabilized silica dispersions II. Experiments on phase separation induced by temperature variation. J. Colloid Interface Sci. **114**, 481–491 (1986)
42. J. Moonen, A. Vrij, Determination of the structure factor for concentrated silica dispersions using small angle x-ray scattering. Colloid Polym. Sci. **266**, 1140–1149 (1988)
43. H. Verduin, J.K.G. Dhont, Phase-diagram of a model adhesive hard-sphere dispersion. J. Colloid Interface Sci. **172**, 425–437 (1995)
44. P.W. Rouw, C.G. de Kruif, Adhesive hard-sphere colloidal dispersions: fractal structures and fractal growth in silica dispersions. Phys. Rev. A **39**, 5399–5408 (1989)
45. A.P.R. Eberle, N.J. Wagner, B. Akgun, S.K. Satija, Temperature-dependent nanostructure of an end-tethered octadecane brush in tetradecane and nanoparticle phase behavior. Langmuir **6**, 3003–3007 (2010)
46. M. Sztucki, T. Narayanan, G. Belina, A. Moussaïd, F. Pignon, H. Hoekstra, Kinetic arrest and glass-glass transition in short-ranged attractive colloids. Phys. Rev. E **74**, 051504-1–051504-12 (2006)
47. W. Rasband. ImageJ. http://rsb.info.nih.gov/ij (2011)
48. Origin. http://www.originlab.com (2011)
49. D.R. Lide (ed.), *CRC Handbook of Chemisty and Physics* (CRC Press, Boca Raton, 2008)
50. B.A. Korgel, S. Fullam, S. Connolly, D. Fitzmaurice, Assembly and self-organization of silver nanocrystal superlattices: ordered "Soft Spheres". J. Phys. Chem. B **102**, 8379–8388 (1998)
51. R.G. Nuzzo, L.H. Dubois, D.L. Allara, Fundamental studies of microscopic wetting on organic surfaces. 1. formation and structral characterization of a self-consistend series of polyfunctional organic monolayers. J. Am. Chem. Soc. **112**, 558–569 (1990)
52. K.A. Dill, S. Bromberg, Molcular driving forces: statistical thermodynamics in chemistry and biology. (Garland Science, New York 2002)
53. A. Badia, S. Singh, L. Demers, L. Cuccia, G.R. Brown, R.B. Lennox, Self-assembled monolayers on gold nanoparticles. Chem. A Eur. J. **2**, 359–363 (1996)
54. A. Ulman, Formation and structure of self-assembled monolayers. Chem. Rev. **96**, 1533–1554 (1996)
55. A. Philipse, A. Vrij, Preparation and properties of nonaqueous model dispersions of chemically modified, charged silica spheres. J. Colloid Interface Sci. **128**, 121–127 (1989)
56. J.W. Cahn, Phase separation by spinodal decomposition in isotropic systems. J Chem. Phys. **42**, 93–99 (1965)
57. D. Nykypanchuk, M.M. Maye, D. van der Lelie, O. Gang, DNA-guided crystallization of colloidal nanoparticles. Nature **451**, 549–552 (2008)
58. G. Foffi, C. de Michele, F. Sciortino, P. Tartaglia. Scaling of dynamics with the range of interaction in short-range attractive colloids. Phys. Rev. Lett. **94**, 078301-1–078301-4 (2005)
59. K.A. Dawson, The glass paradigm for colloidal glasses, gels, and another arrested states driven by attractive interactions. Curr. Opin. Colloid Interface Sci. **7**, 218–227 (2002)

60. G. Foffi, G.D. McCullagh, A. Lawlor, E. Zaccarelli, K.A. Dawson, F. Sciortino, P. Tartaglia, D. Pini, G. Stell, Phase equilibria and glass transition in colloidal systems with short-ranged attractive interactions: application to protein crystallization. Phys. Rev. E 65, 031407-1–031407-17 (2002)
61. T. Narayanan, Synchrotron small-angle X-Ray scattering studies of colloidal suspension. In *Applications of Synchrotron Light to Scattering and Diffraction in Materials and Life Sciences*, ed. by T.A. Ezquerra, M. Garcia-Gutierrez, A. Nogales, M. Gomez. (Springer, Berlin, 2009), pp. 133–156
62. M. Kotlarchyk, S.-H. Chen, Analysis of small angle neutron scattering spectra from polydisperse interacting colloids. J. Chem. Phys. **79**, 2461–2469 (1983)
63. J.L. Yarnell, M.J. Katz, R.G. Wenzel, S.H. Koenig, Structure factor and radial distribution function for liquid argon at 85K. Phys. Rev. A **7**, 2130–2144 (1973)
64. L. Rayleigh, The incidence of light upon a transparent sphere of dimensions comparable with the wave-length. Proc. R. Soc. A **84**, 25–64 (1910)
65. L.A. Feigin D.I. Svergun, *Structure Analysis by Small-Angle X-Ray and Neutron Scattering* (Plenum Press, New York, 1987)
66. M. Sztucki. SAX Sutilities. http://www.sztucki.de/SAXSutilities/ (2011)
67. B.J. Berne, R. Pecora, *Dynamic Light Scattering with Applications to Chemistry, Biology and Physics*. (Wiley, New York, 2000)
68. D.E. Koppel, Analysis of macromolecular polydispersity in intensity correlation spectroscopy: the method of cumulants. J. Chem. Phys. **57**, 4814–4820 (1972)
69. J.W. Virden, J.C. Berg, The use of photon corelation spectroscopy for estimating the rate constant for doublet formation in an aggregating colloidal dispersion. J. Colloid Interface Sci. **149**, 528–535 (1992)
70. H. Holthoff, S.U. Egelhaaf, M. Borkovec, P. Schurtenberger, H. Sticher, Coagulation rate measurements of colloidal particles by simultaneous static and dynamic light scattering. Langmuir **12**, 5541–5549 (1996)
71. H. Holthoff, A. Schmitt, A. Fernandez-Barbero, M. Borkovec, M.A. Cabrerizo-Vilchez, P. Schurtenberger, R. Hidalgo-Allvarez, Measurement of absolute coagulation rate constants for colloidal particles: comparison of single and multiparticle light scattering techniques. J. Colloid Interface Sci. **192**, 463–470 (1997)
72. D.A. Weitz, J.S. Huang, M.Y. Lin, J. Sung, Limits of the fractal dimension for irreversible kinetic aggregation of gold colloids. Phys. Rev. Lett. **54**, 1416–1419 (1985)
73. D.A. Weitz, M.Y. Lin, Dynamic scaling of cluster-mass distribution in kinetic colloid aggregation. Phys. Rev. Lett. **57**, 2037–2040 (1986)
74. P. Dimon, S.K. Sinha, D.A. Weitz, C.R. Safinya, G.S. Smith, W.A. Varady, H.M. Lindsay, Structure of aggregated gold colloids. Phys. Rev. Lett. **57**, 595–598 (1986)

Chapter 5
Origin of Order in Alkyl Thiol-Stabilized Nanoparticle Packings

Abstract The self-assembly of colloidal nanoparticles into regular, crystalline lattices requires the transition from a dispersed fluid state to close-packed solid state of the particles. In charge-stabilized suspensions such a transition leads to the formation of irregular packings. In this chapter we describe a mechanism that allows sterically stabilized nanoparticles to form regular packings. The main requirement is a high mobility of the particles in the agglomerate, allowing them to diffuse and to reach crystalline lattice sites. Mobility is provided in the suspensions used here by a fluid lubricant ligand layer between the particles. This makes the crystalline agglomeration a function of the melting point of the ligand layer of the particles.

5.1 Introduction

The self-assembly of sterically stabilized nanoparticles into regular functional materials also known as colloidal crystals has left the state of being an academic curio. Standardized methods for the production of colloidal crystals by precipitation or evaporation have been developed [1–4]. Research has advanced to explore collective electronic, optical or mechanical properties of the assemblies [5–9]. This fact is in stark contrast to the little consensus on the relation between the agglomeration mechanism and the colloidal crystallization. The colloidal crystallizations in sterically stabilized nanoparticle suspensions have been attributed to: an entropic WAINWRIGHT- ALDER transition of hard spheres upon concentration increase [1, 10], a crystallization of spherical particles under the action of VAN DER WAALS-attraction among particle cores [11, 12], an effect of alignment of attracting ligand bundles on the particle surfaces [13, 14] or an effect of the alignment of particle

Reprinted excerpt with permission from T. Geyer, P. Born, and T. Kraus. Phys. Rev. Lett. 109, 128302 (2012). Copyright 2013 by the American Physical Society. The original article is available online: http://link.aps.org/doi/10.1103/PhysRevLett.109.128302.

P. G. Born, *Crystallization of Nanoscaled Colloids*, Springer Theses,
DOI: 10.1007/978-3-319-00230-9_5, © Springer International Publishing Switzerland 2013

facets [15, 16]. These arguments are based on the assumption that the particles take and reach their thermodynamic equilibrium. Under these preassumption all particle suspensions with low size dispersity should exhibit crystallization upon agglomeration.

However, suspensions of stearyl alcohol-stabilized silica particle suspensions and alkyl thiol-stabilized gold particle suspensions have shown to form amorphous agglomerates [17–22] (see Chap. 4). In these experiments agglomeration was induced by cooling the suspension, indicating an attraction among the particles, and the used particles were narrowly distributed in size [23, 24]. The absence of crystallization in these suspension demonstrates that thermal equilibrium arguments are incomplete for describing colloidal crystallization of sterically stabilized nanoparticles.

Conclusions on mechanisms hindering formation of colloidal crystals might be drawn from theories on growth of atomic clusters, despite some limitations in the analogy between atoms and colloids. Deviations of atomic cluster structures from predicted thermal equilibrium structures has led to the inclusion of the growth process in cluster formation theories. Two models are used in simulations on cluster growth: solid-like growth and liquid-like growth [25]. In both models the clusters grow by the addition of single atoms.

In liquid-like growth the cluster solidifies after its growth is completed, that is, in a further cooling stage. In this case, the final cluster structure does not depend on the details of the growth process, but rather on the cooling process. Depending on the cooling rate, metastable configurations can be solidified [25]. The suppression of the liquid phase in colloidal agglomerates for interaction ranges small compared to the particle diameter [26–28] render the liquid-like growth model unlikely to occur in colloidal suspensions. However, liquid-like behavior should still be considered for small nanoparticles. A mobilization of nanoparticles by suitable solvent vapor is reported [29, 30]. Thus the particles may assume a highly mobile phase in presence of solvent, and solidify only upon full evaporation of solvent.

In the second cluster-growth model, solid-like growth, the growing cluster takes a static structure. The number and stability of certain adsorption sites define the growth process [25]. DIXIT and ZUKOSKI have developed a local description of a solid-like growth model for colloids [31]. In their description the formed agglomerates take a static structure. Agglomerates grow by the adsorption of single particles after BROWNIAN encounters. After adsorption to the agglomerate, the particle can perform three processes. First, it can desorb to become a free particle again. Second, it can diffuse on the agglomerate surface until it finds a crystalline lattice site. And third, another particle can adsorb to the first one what hinders the surface diffusion. The last process would cause the formation of a locally amorphous packing.

Lowering particle concentrations reduces the probability of two-particle collisions on the agglomerate surface. According to DIXIT'S and ZUKOSKI'S model, sufficient reduction of the concentration guarantees the formation of colloidal crystals upon agglomeration. Only if additional mechanism prevent diffusion on the agglomerate surface, or at least slow down surface diffusion so strongly that for all practical particle concentration two-particle collision occur on the agglomerate surface, agglomeration may form amorphous packings.

The colloidal crystallization of alkyl thiol-stabilized gold nanoparticles is investigated in this chapter with emphasis on the mobility of the particles on the agglomerates following above considerations. The particles are characterized using transmission electron microscopy and small angle X-ray scattering, proving a low size dispersity of the particles. The mobility of the particles is expected to depend on the temperature. In precipitation experiments the influence of the temperature and the ligand chain length on the crystallization behavior is evaluated. A distinct minimal temperature for the formation of colloidal crystals is found. Using differential scanning calorimetry and a variation of diffusing-wave spectroscopy a steep increase in particle mobility on the agglomerate surface at the melting point of the ligand shell is found. This increase in mobility coincides with the onset of colloidal crystallization. A liquid-like ligand shell thus lubricates particle diffusion and allows formation of colloidal crystals.

5.2 Experimental

The experiments seek to clarify mechanisms preventing or enforcing formation of colloidal crystals in sterically stabilized particles. The influence of temperature is tested by precipitation experiments at various temperatures. The formation of colloidal crystals is compared to measurements on the interactions among the ligands and mobility measurements.

5.2.1 Particle Synthesis and Characterization

Gold nanoparticles with dodecyl thiol (C12), hexadecyl thiol (C16) and octadecyl thiol (C18) ligands suspended in heptane were available from the experiments described in Chap. 4. For agglomeration experiments 1-propanol was used as incompatible solvent. 1-propanol was chosen as it mixes well with heptane and has a matching boiling point. This minimizes changes in the solvent constitution during evaporation of the solvent for electron microscopy evaluations.

The metal cores of the particles were characterized with transmission electron microscopy (TEM, Philips CM200, 200 keV) and small-angle X-ray scattering (SAXS, beamline BW4 of DORIS III at DESY, Hamburg, $\lambda = 0.138$ nm).

The calorimetric behavior of the ligands was investigated with differential scanning calorimetry (DSC) using a Mettler Toledo DSC822e. Approximately 20 ml of suspension were centrifuged and dried to form dense pellets of particles with a mass of $\approx 8\,\mu$g. The pellets were transferred to open alumina crucibles. The energy uptake was measured against an empty crucible as standard in a nitrogen stream with a scan rate of 10 °C/min.

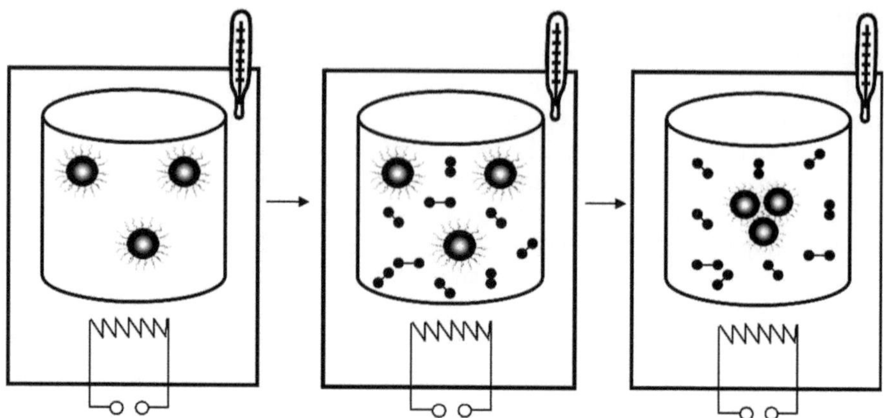

Fig. 5.1 Scheme of the precipitation experiments. The suspensions were equilibrated at the target temperature, the alcohol was added, and finally the structures of the formed agglomerates were investigated

5.2.2 Precipitation Experiments

Agglomeration experiments followed the scheme depicted in Fig. 5.1. The particle suspensions were homogenized at elevated temperatures and subsequently pressed through syringe filters that had been heated to 45 °C in an oven. Without heating the filter the suspension would be quenched below the agglomeration temperature for the longer akyl ligands resulting in a strong particle capture by the filter membrane. The suspensions were diluted to particle concentrations of $\approx 10^{11}$ ml^{-1}. The suspensions and 1-propanol were separately thermally equilibrated in a Eppendorf Thermomixer Comfort. After equilibration at the respective target temperature the suspension and the alcohol were mixed in a 1:1 ratio and let rest for two hours at the respective target temperature. The formed sediments were collected, pipetted onto a TEM grid and left to dry. For SAXS measurements the same route was applied, with the suspension contained in capillaries in a thermostated water bath. After resting for two hours, scattering measurements were performed on the formed sediment. These agglomeration experiments were performed at temperatures starting from -20 °C in 5 °C steps up to 60 °C.

5.2.3 Mobility Measurements

A conventional setup for dynamic light scattering (DLS) was used (ALV CGS-3, $\lambda = 633$ nm) for measurements of the mobility of C16 and C18 particles on agglomerate surfaces. The setup is equipped with a cuvette rotation unit for non-ergodic samples and an index matching, thermostated solvent bath for temperature control.

Roughly 25 ml of suspension were centrifuged and dried to form dense pellets of particles with an mass of $\approx 10\,\mu g$. The pellets were transferred to the cylindrical DLS glas cuvettes. The pellets were resuspended in 0.2 ml heptane using an ultrasonic bath. The dense suspensions were rapidly dried under reduced pressure of 30 mbar while constantly rotating the horizontal cuvette. This procedure yielded porous metallic-golden particle films with amorphous particle packings on the inside of the glass cuvettes. The cuvettes were subsequently mounted in the cuvette rotation unit rotating at 7.5 rpm, leading to an echoed scattering signal. The scattered light was recorded at an angle of 145° for 24 h with a time resolution of $102.4\,\mu s$. The autocorrelation of the time trace of the scattered light was fit with an exponential decay to quantify the particles' mobility. Measurements were performed at temperatures from 30 to 60 °C.

Details of the experimental approach and the data analysis can be found in the appendix of this chapter.

5.3 Results

In the following, we characterize the particles and investigate the formation of ordered colloidal crystal upon precipitation. We find a ligand length-dependent minimal temperature to form colloidal crystals. This transition temperature corresponds to melting transitions in the particles' ligand shell and to an increased particle mobility. A restriction of the motion of the particles by solid-like ligand shells can be assumed.

5.3.1 Particle Characterization

Results of the particle characterization can be found in Chap. 4. In summary, gold nanoparticles with a core radius of 3.2 nm and low size dispersity were produced. The use of C12, C16 and C18 thiols lead to corresponding increase in ligand shell thicknesses.

DSC measurements revealed endothermal reactions in the nanoparticle pellets for the C16 and the C18 sample (Fig. 5.2). The onset of the reactions were ≈ 40 °C for the C16 sample and ≈ 50 °C for the C18 sample, and the reactions continued for ≈ 20 °C. No reaction was detectable in the C12 sample.

5.3.2 Precipitation Experiments

The TEM investigations revealed a remarkable dependency of the formation of crystalline packings in the agglomerates on temperature and ligand chain length (Fig. 5.3). The transition from amorphous packings to crystalline packings occured between

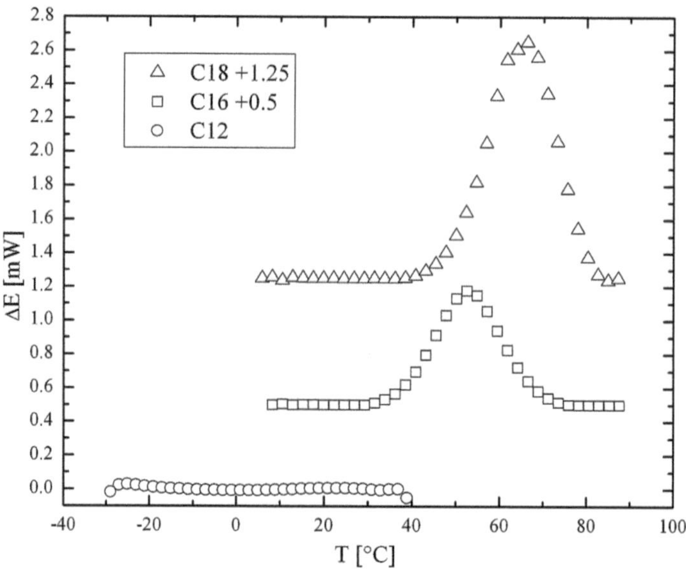

Fig. 5.2 DSC curves of C12-C18 particles. The curves are offset by the indicated amounts. An endothermal reaction in the C16 and the C18 sample is detected by an energy consumption ΔE of the samples, indicating a melting of the ligand shell

-10 and $-5\,°C$ for the C12 sample, between 40 and $45\,°C$ for the C16 sample, and between 50 and $55\,°C$ for the C18 sample.

The morphologies of the crystallites varied considerably within a batch and a precipitation experiment. Morphologies including hexagons, truncated triangles, slabs and spheres are observed above the onset of the regular packings Fig. 5.4. A reason might be that the nucleation of agglomerates and the adsorption kinetics of free particles were not well controlled in the experiments. Agglomerates that had nucleated at early times experienced a much higher free particle concentration than agglomerates formed at later stage, which also had less time to rearrange to find the morphology with the smallest free energy.

SAXS measurements show a minimal temperature for the formation of crystalline packings, too. Figure 5.5 shows $S(q)$-curves obtained from sediments formed at the respective highest temperature where no regular packing were observed and the lowest temperature at which TEM indicated regular packings. At temperatures without observable regular packings, $S(q)$ shows only the wavy shape of irregular fluids or amorphous solids. At temperatures with regular packings, sharp intense peaks evolve in $S(q)$. However, a clear indexing of the peaks to a specific crystalline lattice is not possible. The chosen precipitation route provided a very rapid agglomeration within minutes. This apparently results not only in a variety of morphologies but also in varying internal packing structures explaining the various peaks.

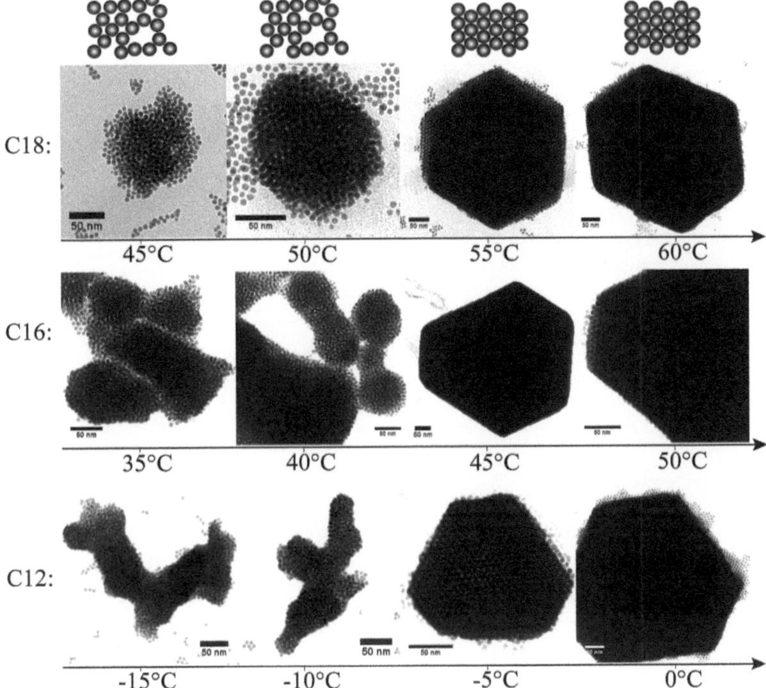

Fig. 5.3 Electron micrograph series of agglomerates grown at the temperatures given below. Crystalline packings were only observed at the temperatures where hexagonal agglomerates are displayed. The temperatures at which crystalline packings are observed increase with ligand chain length. All scale bars correspond to 50 nm

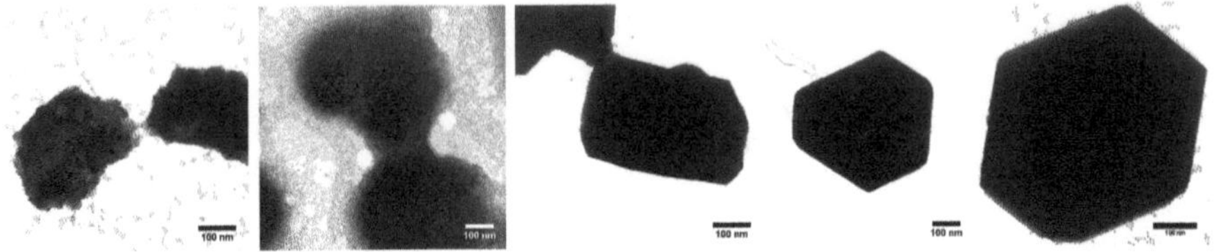

Fig. 5.4 A provisional classification of crystalline agglomerate morphologies. The variety of morphologies, from irregular, globular, slabs, truncated triangles to hexagons, stem from probably varying agglomeration kinetics experienced by the agglomerates

5.3.3 Mobility Measurements

In the introduction of this chapter a link between the mobility of the particles in agglomerate surfaces and the possibility to form crystalline particle packings is established. We quantify this mobility with the decay of the autocorrelation of scattered light from particle films. Due to the low penetration depth of light and high reflectivity

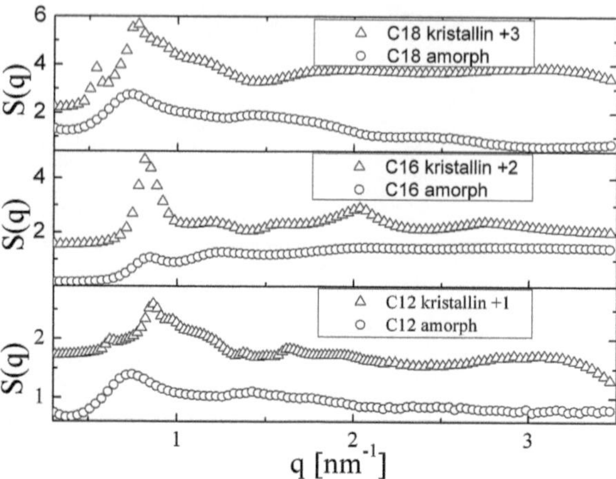

Fig. 5.5 S(q) curves obtained from scattering from sediments formed at the given temperatures. Curves at the respective higher temperature are shifted by the denoted amount. Sharp BRAGG peaks are only observable at elevated temperatures, proving the suppression of crystalline packings at lower temperatures

of metals this method is very sensitive to surface reconstructions of the films, which stem from motion of the particles. The autocorrelation of the scattered light signal is fit using an exponential decay. The decay constant exhibits a strong increase at temperatures coinciding with the melting of the ligand shell and the temperatures needed for forming crystalline agglomerates (Fig. 5.6). The increase is more pronounced for the C18 particles, where it increases from $1.8 \cdot 10^{-7}$ s^{-1} at 30° to $3.1 \cdot 10^{-6}$ s^{-1} at 60°. The increase measured with C16 particles from $3.6 \cdot 10^{-7}$ s^{-1} at 30° to $2.2 \cdot 10^{-6}$ s^{-1} at 60° is still more than 500 %.

5.4 Discussion

TEM and SAXS measurements confirmed uniform particle cores for all syntheses. Differences in properties of the suspensions must therefore arise from the varying length of the ligand shells.

The DSC measurements revealed an endothermal reaction depending on the chain length of the ligands in the nanoparticle pellets. This reaction can be attributed to a melting transition in the ligand shell of the nanoparticles [32]. A comprehensive study of melting transitions of alkyl thiol ligands with 12–20 C-atoms on gold nanoparticles was performed by BADIA et al. [33]. They report a melting transition for C12 samples beginning at $-20\,°$C, albeit 5 times less intense than the transition in their C18 sample. Our instrument could probably not detect this existing transition in our C12 sample, because its resolution was insufficient. The reported melting temperatures of the C16 and the C18 samples in BADIA's study are shifted by $10\,°$C to lower temperatures compared to our results. This might be explained by the smaller metal cores of the particles used in their experiments, which exhibit 1.5 nm average radius compared to 3.2 nm average radius of our particles. The greater surface curvature of the smaller

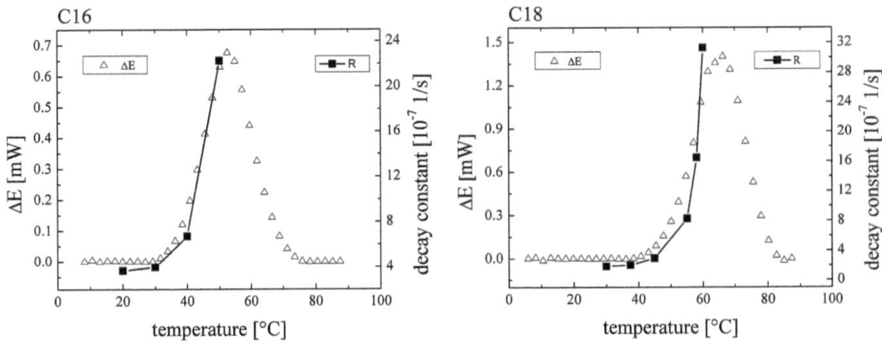

Fig. 5.6 Results of light scattering measurements on a solid C16 (*left*) and a solid C18 (*right*) sample. For aid of interpretation, the energy consumption ΔE measured with DSC of the respective samples are plotted in the same graph. The decay of the autocorrelation function R, indicator for the particle mobility, increases steeply with increasing the temperature. The increase in mobility coincides with the onset of melting of the ligand shell

particles might reduce interactions among the chains and induce stresses in the ligand shell and, consequently, lower the melting point (which is also consistent with the high melting points above $100\,^\circ$C of self-assembled alkyl thiol monolayer on planar gold [34]). From this considerations an onset of melting of the ligand shell of our C12 particles can be expected around $-10\,^\circ$C.

The onset of crystallization of particle agglomerates coincides with the onset of melting of the ligand shell (Table 5.1). The longer the alkyl chain is, the higher are melting temperature and crystallization temperature. DSC measurements of ligand shell melting rather than bulk alkane melting thus allows prediction of temperatures needed to produce colloidal crystals.

The link between the DSC results and the crystallization behavior of the agglomerates is given by the light scattering results. The mobility of the particles increase by more than one order of magnitude at the melting temperature of the ligands. The solid-like ligand chains on the particles most likely form bundles, which align and coalesce upon contact of the particles, and restrict further motion of the particles. Above the melting temperature, the ligands form viscous layer around the particles, which still can coalesce upon particle contact and only slow down but do not prevent the diffusion of particles.

Table 5.1 Temperatures of onset of melting of alkyl thiol ligands T_m and of formation of colloidal crystals T_c

Chain length	T_m alkyl thiol ligands [°C]	T_c gold nano-particles [°C]
C12	≈ -10[a]	≈ -7.5
C16	≈ 40	≈ 42.5
C18	≈ 50	≈ 52.2

[a] The melting temperature of the C12 ligands is estimated on the basis of results in literature [33], see text for details

In summary, an adjusted growth model for sterically stabilized particles can be obtained. According to the SAXS results the particles obtain their crystalline structure already in the suspension, a liquid-like growth model for the agglomerates thus can be ruled out. However, the growth is fully solid-like only at temperatures below the melting point of the ligands on the particles. In this case the adsorption of a particle on the agglomerate surface can be visualized as a solid-solid sintering of the ligand shells, leaving the particle immobilized on the agglomerate surface. At temperatures above the melting point the adsorption process is rather a liquid-liquid coalescence of the molten ligand-shells. The entanglement of chains may slow down the particles, but does not prevent "sliding" and "rolling" of the particles. Compared to a direct contact of the core surfaces or the contact of solid-like ligand shells, the motion is essentially lubricated. The particles can perform surface diffusion and arrange into a crystalline lattice after an BROWNIAN encounter. This lubrication effect of the ligand chains becomes obvious when comparing to the well-investigated agglomeration of charge-stabilized colloids. The agglomeration in these suspensions varies from fast diffusion-limited agglomeration within minutes to very slow reaction-limited agglomeration on the order of days or weeks. Still, in all agglomeration regimes, low-density ramified agglomerates with varying fractal dimensions are formed [35–42]. The development of models for formation of amorphous packings in colloidal suspensions should be possible in analogy to models for granular matter. In these macroscopic many-particle systems friction among the particles is often governing the packing [43–48].

The precipitation experiments shown here were performed at low particle concentrations of $n \approx 10^{11}$ ml^{-1}. An initial collision rate $J = 4kT/3\eta \cdot n \approx 1$ s^{-1} can be estimated from Smoluchowski's theory [49]. η denotes the solvent viscosity, kT the thermal energy. A negligible influence of many-particle effects on the crystallization can be expected at this low collision rate. Further experiments should evaluate whether increasing the particle concentration delays the onset of crystallization to higher temperatures because higher mobilities are needed to ensure a higher chance of the particles to diffuse into a crystalline site than being captured by a second particle. It may be that with liquid ligand shells the mobility is sufficient to have surface diffusion even on crowded surfaces, and the particles can rearrange and crystallize at particle concentrations up to the hard-sphere glass transition limit [50, 51].

Further, the role of the solvent is not yet clear. The melting transitions of the ligands were determined by DSC on dried samples in air. The increase of mobility of the particles also was measured by light scattering in air. The precipitation experiments were performed in suspension. Stability analysis of alkyl thiol-stabilized gold nanoparticle suspensions indicated molten ligand shells down to the freezing temperature of the respective bulk alkane (Chap. 4). Thus following above argumentation, crystalline agglomerates should be produced upon precipitation down to the freezing temperature of the respective bulk alkane. We believe that this depression of the melting temperatures of the ligand shells from the measured melting temperatures in air down to the alkane freezing temperature in suspension is present only in good solvents like heptane. Addition of the alcohol lowers solvent quality, raises the freezing temperature and causes precipitation of the unpolar particles.

5.5 Conclusion

In precipitation experiments, the formation of crystalline agglomerates of alkyl thiol-stabilized gold nanoparticles coincides with the melting temperature of the ligands of the particles. We explain this behavior with the ligand shell forming a liquid lubricant layer for the rearrangement of the particles. Upon contact, the particles ligand shells merge, but do not prevent rearrangement of the particles and reaching the crystalline lattice sites. The formation of crystalline packings even under rapid agglomeration of the particles by precipitation indicate an inferior role of the agglomeration kinetics on the formation of crystalline packings.

5.6 Appendix

5.6.1 Mobility Measurements Using Light Scattering

Dynamic light scattering as presented in the appendix of Chap. 4 is an ideal method to gain information on the motion of particles. The method averages a great number of particles in suspension, is non-destructive and applicable *in-situ*. However, when applied to the diffusion of particles on agglomerate surfaces, three problems arise. First, a typical assumption in the evaluation of dynamic light scattering data is that every photon experiences only a single scattering event, as is the case in dilute suspensions but not in agglomerates. Second, light scattering probes length scales on the order of the wavelength of light. In surface diffusion particle mobility may be very low and displacements over hundreds of nanometer may take very long or never occur. Finally, the diffusion process may be non-ergodic, becuase not all particles experience the same particle film surface morphology and consequently the same mobility. The measurement thus depends on the volume of the sample probed during the measurement.

To make dynamic light scattering applicable to dense particle suspensions, the evaluation of the scattered light signal has been extended to account for multiple scattering in the suspension by a technique termed diffusing wave spectroscopy (DWS) [52]. It is assumed that the photons diffuse through the sample. The autocorrelation (Eq. 4.17) thus turns into a product of the i independent correlation functions of i scattering events in the sample averaged over scattering vector q so as to reflect the average scattering event in the path:

$$C(q,\tau)^i = \langle n(q,t)n(q,t+\tau)\rangle_q^i .$$ (5.1)

This diffusing wave approach enables evaluation of data gained from samples with high concentrations. Contributions by small rearrangements of particles accumulate due to the multiple scattering of a photon and the method becomes sensitive even to motion on scales much smaller than the wavelength of the incident light. However, the method only partially solves the problem of non-ergodic samples: more particles are probed compared to conventional DLS, but the probed sample volume is still limited to the illuminated volume. Also, fast processes such as dust particles diffusing in the pathway of the illuminating laser beam or air bubbles in the thermostating solvent bath overlay the autocorrelation of slow particle surface diffusion, creating problems determining the contribution of the particles.

A method developed to render DLS applicable to non-ergodic samples and minimize unwanted contributions of dust and bubbles is echoed dynamic light scattering (eDLS) [53]. In this method, the sample is rotated at a constant velocity while recording the scattered light. After each rotation the detector 'sees' the same sample volume again. The autocorrelation of the scattered light thus exhibits peaks termed 'echoes' in correlation at every integer multiple of the rotation period. A slow decay in correlation due to particle rearrangement causes a decay of the envelope of the peaks. Processes with correlation times shorter than the rotation period of the cuvette are not detected. With little additional computational work to filter the envelope of the peaks, this method allows suppression of dust contributions to the autocorrelation and multiplies the probed volume to gain an ensemble average of the sample. Imperfections in the rotation are shown to cause a broadening of the peaks, but for fluctuations in angular speed and wobbling of the cuvette about a fixed mean the envelope reproduces the slow decay due to particle rearrangements.

Combined, DWS and eDLS should provide the tools to measure particle mobilities on particle agglomerate surfaces. However, two problems of sample preparation arise. The agglomerates have to be fixed such that settling or diffusion of the agglomerates in suspension do not cause a decay of the autocorrelation. The limited penetration of light into metals ensures a sensitivity of the scattered light signal to surface diffusion of particles rather than bulk diffusion, but hinders transmission measurement on bulk particle agglomerate sediments as used in conventional DWS measurements. Sample preparation thus has to ensure an exposure of particle agglomerate surfaces to the incident laser light in a geometry suited to cause multiple scattering between surfaces.

We approached this problems as described in Sect. 5.2.3. A slurry of alkyl thiol-stabilized gold nanoparticles was produced in the DLS cuvette, from which the solvent was rapidly evaporated under constant rotation of the cuvette. This procedure lead to macroscopically homogeneous golden particle films on the inner side of the cuvette (Fig. 5.7). On a microscopic scale, the particles agglomerated to form sponge-like structures (Fig. 5.8). We assume that light enters the pores and is trapped, i.e. scattered multiple times prior to leaving the sample, allowing to detect small rearrangements.

The echoed signal was produced by rotation of the cuvette. The scattered light was detected at an angle of 145°. The autocorrelation of the signal and the envelope to the correlation peaks was computed using our own Matlab [54]-algorithm. An exponential decay was fit to the envelope using Origin [55]. For calculation of diffusion coefficients the value of the transport mean free path of the photons, l^*, and the thickness of the sample that the photons have traveled through, L, is needed [52].

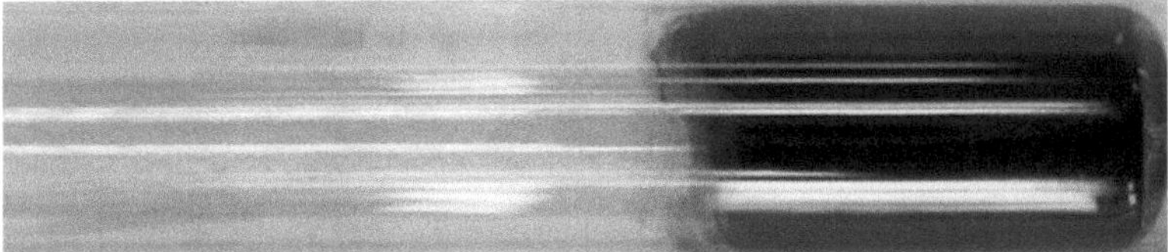

Fig. 5.7 Macroscopic appearance of the particle films inside the DLS-cuvettes used for mobility measurements in the solid phase

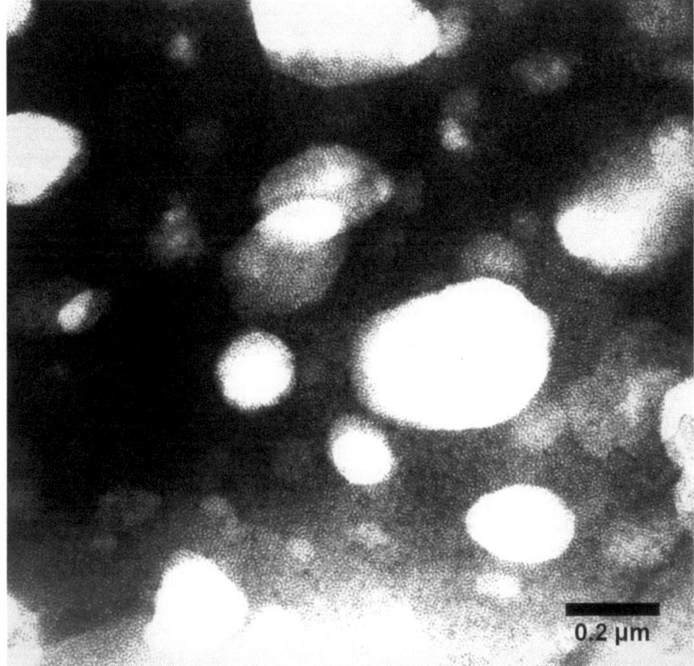

Fig. 5.8 Microscopic appearance of the particle films used for mobility measurements in the solid phase

These values are presently unknown; thus only the decay constants as a measure of the mobility are presented.

References

1. M.I. Bodnarchuk, L. Li, A. Fok, S. Nachtergaele, R.F. Ismagilov, D.V. Talapin, Three-dimensional nanocrystal superlattices grown in nanoliter microfluidic plugs. J. Am. Chem. Soc. **133**, 8956-8960 (2011)
2. D.V. Talapin, E.V. Shevchenko, A. Kornowski, N. Gaponik, M. Haase, A.L. Rogach, H. Weller, A New Approach to Crystallization of CdSe Nanoparticles into Ordered Three-Dimensional Superlattices. Adv. Mater. **13**, 1868–1871 (2001)

3. T.P. Bigioni, X.-M. Lin, T.T. Nguyen, E.I. Corwin, T.A. Witten, H.M. Jaeger, Kinetically driven self assembly of highly ordered nanoparticlemonolayers. Nat. Mater. **5**, 265–270 (2006)

4. C.B. Murray, C.R. Kagan, M.G. Bawendi, Synthesis and Characterization of Monodisperse Nanocrystals and Close-Packed Nanocrystal Assemblies. Ann. Rev. Mater. Sci. **30**, 545–610 (2000)

5. S.-H. Kim, G. Medeiros-Ribeiro, D.A.A. Ohlberg, R.S. Williams, J.R. Heath, Individual and collective electronic properties of Ag Nanocrystals. J. Phys. Chem. **103**, 10341–10347 (1999)

6. M.-P. Pileni, Self-assembly of Inorganic Nanocrystals: Fabrication and collective intrinsic properties. Acc. Chem. Res. **40**, 685–693 (2007)

7. P. Podsiadlo, B. Lee, V.B. Prakapenka, G.V. Krylova, R.D. Schaller, A. Demortiere, E.V. Shevchenko, High-pressure structural stability and elasticity of supercrystals self-assembled from Nanocrystals. Nano Lett. **11**, 579–588, (2011)

8. J. He, P.K.N.L. Frazer, A. Weis, X.-M. Lin, H.M. Jaeger, Fabrication and mechanical properties of large-scale freestanding nanoparticle membranes. Small **6**, 1449–1456 (2010)

9. D.V. Talapin, J.-S. Lee, M.V. Kovalenko, E. Shevchenko, Prospects of colloidal nanocrystals for electronic and optoelectronic applications. Chem. Rev. **110**, 389–458 (2010)

10. B.A. Korgel, D. Fitzmaurice, Small-angle x-ray-scattering study of silver-nanocrystal disorder-order phase transitions. Phys. Rev. **59**, 14191–14201 (1999)

11. B. Abecassis, F. Testard, O. Spalla, Gold nanoparticle superlattice crystallization probed in situ. Phys. Rev. Lett. **100**, 115504–1–115504–4, (2008)

12. M.B. Sigman, A.E. Saunders, B.A. Korgel, Metal nanocrystal superlattice nucleation and growth. Langmuir **20**, 978–983 (2004)

13. Z.L. Wang, S.A. Harfenist, R.L. Whetten, J. Bentley, N.D. Evans, Bundling and interdigitation of adsorbed thiolate groups in self-assembled nanocrystal superlattices. J. Phys. Chem. **102**, 3068–3072 (1998)

14. Z.L. Wang, S.A. Harfenist, I. Vezmar, R.L. Whetten, J. Bentley, N.D. Evans, K.B. Alexander, Superlattices of self-assembled tetrahedral Ag Nanocrystals. Adv. Mater. **10**, 808–812 (1998)

15. B.A. Korgel, S. Fullam, S. Connolly, D. Fitzmaurice, Assembly and self-organization of silver nanocrystal superlattices: Ordered "Soft Spheres". J. Phys. Chem. **102**, 8379–8388 (1998)

16. S. Disch, E. Wetterskog, R.P. Hermann, G. Salazar-Alvarez, P. Busch, T. Bruckel, L. Bergstrom, S. Kamali, Shape induced symmetry in self-assembled mesocrystals of Iron Oxide Nanocubes. Nano Lett. **11**, 1651–1656 (2011)

17. S. Roke, O. Berg, J. Buitenhuis, A. van Blaaderen, M. Bonn, Surface molecular view of colloidal gelation. Proc. Natl. Acad. Sci. **103**, 13310–13314 (2006)

18. P.W. Rouw, C.G. de Kruif, Adhesive hard-sphere colloidal dispersions: Fractal structures and fractal growth in silica dispersions. Phys. Rev. **39**, 5399–5408 (1989)

19. M. Sztucki, T. Narayanan, G. Belina, A. Moussaïd, F. Pignon, H. Hoekstra, Kinetic arrest and glass-glass transition in short-ranged attractive colloids. Phys. Rev. **74**, 051504–1–051504–12, (2006)

20. J.W. Jansen, C.G. Kruif, A. Vrij, Phase seperation of sterically stabilized colloids as a function of temperature. Chem. Phys. Lett. **107**, 450–453 (1984)

21. J.W. Jansen, C.G. Kruif, A. Vrij, Attractions in sterically stabilized Silica Dispersions I. Theory of phase separation. J. Colloid Interface Sci. **114**, 471–480 (1986)

22. J.W. Jansen, C.G. Kruif, A. Vrij, Attractions in sterically stabilized Silica Dispersions II. Experiments on phase separation induced by temperature variation. J. Colloid Interface Sci. **114**, 481–491 (1986)

23. A.K. van Helden, J.W. Hansen, A. Vrij, Preparation and characterization of spherical monodisperse Silica Dispersions in Nonaqueous Solvents. J. Colloid Interface Sci. **81**, 354–368 (1981)

24. N. Zheng, J. Fan, G.D. Stucky, One-step one-phase synthesis of monodisperse noble-metallic nanoparticles and their colloidal crystals. J. Am. Chem. Soc. **128**, 6550–6551 (2006)

25. F. Baletto, R. Ferrando, Structural properties of nanoclusters: Energetic, thermodynamic, and kinetic effects. Rev. Mod. Phys. **77**, 371–423 (2005)

26. G. Foffi, C. de Michele, F. Sciortino, P. Tartaglia, Scaling of dynamics with the range of interaction in short-range attractive colloids. Phys. Rev. Lett. **94**, 078301–1–078301–4 (2005)

27. K.A. Dawson, The glass paradigm for colloidal glasses, gels, and another arrested states driven by attractive interactions. Curr. Opin. Colloid Interface Sci. **7**, 218–227 (2002)
28. G. Foffi, G.D. McCullagh, A. Lawlor, E. Zaccarelli, K.A. Dawson, F. Sciortino, P. Tartaglia, D. Pini, G. Stell, Phase equilibria and glass transition in colloidal systems with short-ranged attractive interactions: Application to protein crystallization. Phys. Rev. **65**, 031407-1–031407-17 (2002)
29. C. Gutierrez-Wing, P. Santiago, J. Ascencio, A. Camacho, M. Jose-Yacaman, Self-assembling of gold nanoparticles in one, two, and three dimensions. Appl. Phys. A **71**, 237–243 (2000)
30. V. Tarasov, *Control of the Self-Assembly of Alkanethiol-Coated Gold Nanoparticles in the SOlid State* (Master's thesis, Department of Science and Engineering, Masachusetts Institute of Techology , 2008)
31. N.M. Dixit, C.F. Zukoski, Competition between crystallization and gelation: A local description. Phys. Rev. E **67**, 061501-1–061501-13 (2003)
32. A. Badia, W. Gao, S. Singh, L. Demers, L. Cuccia, L. Reven, Structure and chain dynamics of Alkanethiol-Capped Gold Colloids. Langmuir **12**, 1262–1269 (1996)
33. A. Badia, S. Singh, L. Demers, L. Cuccia, G.R. Brown, R.B. Lennox, Self-assembled monolayers on gold nanoparticles. Chem. Europ. J. **2**, 359–363 (1996)
34. F. Schreiber, Structure and growth of self-assembling monolayers. Prog. Surf. Sci. **65**, 151–256 (2000)
35. D.A. Weitz, J.S. Huang, M.Y. Lin, J. Sung, Dynamics of diffusion-limited kinetic aggregation. Phys. Rev. Lett. **53**, 1657–1660 (1984)
36. D.A. Weitz, J.S. Huang, M.Y. Lin, J. Sung, Limits of the fractal dimension for irreversible kinetic aggregation of Gold Colloids. Phys. Rev. Lett. **54**, 1416–1419 (1985)
37. D.A. Weitz, M.Y. Lin, C.J. Sandroff, Colloidal aggregation revisited: New insights based on fractal structure and surface-enhanced raman scattering. Surf. Sci. **158**, 147–164 (1985)
38. D.A. Weitz, M.Y. Lin, Dynamic Scaling of Cluster-Mass Distribution in Kinetic Colloid Aggregation. Physical Review Letters **57**, 2037–2040 (1986)
39. P. Dimon, S.K. Sinha, D.A. Weitz, C.R. Safinya, G.S. Smith, W.A. Varady, H.M. Lindsay, Structure of aggregated Gold Colloids. Phys. Rev. Lett. **57**, 595–598 (1986)
40. M.Y. Lin, H.M. Lindsay, D.A. Weitz, R.C. Ball, R. Klein, P. Meakin, Universality in colloidal aggregation. Nature **339**, 360–362 (1989)
41. M.Y. Lin, H.M. Lindsay, D.A. Weitz, R. Klein, R.C. Ball, P. Meakin, Universal diffusion-limited colloidal aggregation. J. Phys. Condens. Matt. **2**, 3093–3113 (1990)
42. M.Y. Lin, H.M. Lindsay, D.A. Weitz, R.C. Ball, R. Klein, P. Meakin, Universal reaction-limited colloidal aggregation. Phys. Rev. A **41**, 2005–2020 (1990)
43. N. Estrada, E. Azema, F. Radjai, A. Taboada, Identification of rolling resistance as a shape parameter in sheared granular media. Phys. Rev. E, **84**, 011306-1–011306-5 (2011)
44. R.S. Farr, R.D. Groot, Close packing density of polydisperse hard spheres. J. Chem. Phys. **131**, 244104–244104-7 (2009)
45. J.A. Dijksman, E. Wandersman, S. Slotterback, C.R. Berardi, W.D. Updegraff, M. van Hecke, W. Losert, From frictional to viscous behavior: Three-dimensional imaging and rheology of gravitational suspensions. Phys. Rev. E, **82**, 060301-1–060301-4 (2010)
46. C. Song, P. Wang, H.A. Makse, A phase diagram for jammed matter. Nature **453**, 629–632 (2008)
47. C.L. Martin, R.K. Bordia, Influence of adhesion and friction on the geometry of packings of spherical particles. Phys. Rev. E, **77**, 031307-1 - 031307-8 (2008)
48. M. Muthuswamy, A. Tordesillas, How do interparticle contact friction, packing density and degree of polydispersity affect force propagation in particulate assemblies? J. Stat. Mech. Theor. Exp. **2006**, P09003 (2006)
49. D. Myers, Surfaces, interfaces, and colloids: Principles and applications. Wiley, New York, (1999)
50. P.N. Pusey, Liquids, freezing and glass transition. North-Holland (1991)
51. P.N. Pusey, W. van Megen, Phase behavior of concentrated suspensions of nearly hard colloidal spheres. Nature **320**, 340–342 (1986)

52. D.A. Weitz, J.X. Zhu, D.J. Durian, H. Gang, D.J. Pine, Diffusing-wave spectroscopy: The technique and some applications. Physica Scripta **T49**, 610–621 (1993)
53. K.N. Pham, S.U. Egelhaaf, A. Moussaid, P.N. Pusey, Ensemble-averaging in dynamic light scattering by an echo technique. Rev. Sci. Instr. **75**, 2419–2431 (2004)
54. Matlab. http://www.mathworks.com
55. Origin. http://www.originlab.com

Chapter 6
Conclusion and Outlook

The presented work aimed at the formation of high-quality colloidal crystals from nanoparticles and investigated structure formation processes beyond the close-packing. These problems were tackled with two approaches, convective particle assembly and temperature-induced particle assembly. The first approach started with the assembly of micron and sub-micron sized particles and tried to expand the identified mechanisms to nanoparticles. The second approach started with the assembly of nanoparticles. The two used methods also represented two different crystallization principles. The convective assembly was expected to produce dense packings of particles by minimization of free energy under the action of an external potential. The temperature-induced particle assembly was expected to produce dense packings of particles by minimization of free energy under the action of an attractive interparticle potential.

6.1 Conclusion

The convective particle assembly process was optimized to form high quality large-area colloidal crystals in this work. Deposition parameters controlling the formation and spreading of defects in the packing were determined. Under optimized conditions, the formation of crystalline films was only limited by the setup and substrate size. Convective assembly thus could be used in future for masks, templates or functional coatings. The underlying assembly mechanism was found to rely on two mechanisms, one based on drag forces acting on the particles and another one originating from capillary interactions among the particles. The transition between crystallization by convective and by capillary mechanisms depended on particle size and temperature. For temperatures below 20 °C and particles smaller than 100 nm the attractive capillary crystallization dominated. The two mechanisms ensured formation of crystalline packings for all tested particle sizes and temperatures. However, the crystallization mechanisms were found to affect the packing of the particles and

P. G. Born, *Crystallization of Nanoscaled Colloids*, Springer Theses,   121
DOI: 10.1007/978-3-319-00230-9_6, © Springer International Publishing Switzerland 2013

could be used to tune the structure. Convective crystallized particle films exhibit large grains, under optimized conditions up to square millimeters large. Capillary crystallized particle films exhibited small grains only few particle diameters large. These results suggest, that colloidal crystals of nanoparticles can be produced using convective assembly at ambient conditions only by the capillary mechanism.

The temperature-induced particle assembly surprisingly failed to produce colloidal crystals. Although the results indicated attraction among the nanoparticles, no crystalline packings were formed upon agglomeration. To achieve crystals, precipitation of the nanoparticles by polar solvents at elevated temperatures was required. The temperature needed for achieving crystalline packings depended on the melting temperature of the surface ligands of the particles. With molten ligand shells the particles in the close-packing of an agglomerate are lubricated and the particles can diffuse to crystalline lattice sites. The kinetics of the agglomeration process were shown to be either diffusion-limited or reaction-limited, depending on temperature. The kinetics could be used as a structure-directing parameter: both the formation of low-density fractal structures in diffusion-limited agglomeration and the formation of dense globular structures and of agglomerate superstructures in reaction-limited agglomeration could be induced.

The two approaches used are still very different. The first assembles large charge-stabilized particles in water by an external potential into two-dimensional crystals, the second assembles small, unpolar nanoparticles purely by internal attraction into three-dimensional structures. However, temperature acted as a structure-directing parameter in both approaches. The packing and the morphology of the formed particle packings could be adjusted by small temperature variations. In convective assembly, depending on temperature the assembly is dominated by capillary mechanisms or by convective mechanisms. This affects the packing density and the grain size of the formed crystals. In temperature-induced assembly, the temperature controls the morphology of the assemblies. In combination with precipitation by polar solvents, the suspension temperature could even be used to switch between formation of amorphous and crystalline packings. These possibilities to control the assembly processes by temperature are new aspects in colloidal crystallization research. The formation of a specific packing was assumed so far to be a function of the particle shape and interaction. We could show that the same particles in the same assembly approach can be directed into diverse structures. This is an important step towards manufacturing of materials: one particle type with its particle-specific properties can be used to create various structures with different structure-specific properties.

In conclusion, in this work we developed fundamental aspects of both approaches to colloidal crystals, convective particle assembly and temperature-induced particle assembly. The approaches are applicable for production of nanostructures and functional materials. Processes leading to structures and packings beyond hexagonal packings even from spheres were demonstrated.

6.2 Outlook

Functional particles can be arranged using the methods developed here. This requires the identification of useful particles and adapting of the methods investigated in this work to the requirements of applications. The technique developed in this work to produce large-area high quality crystalline particle films by convective assembly is applicable to the production of masks, templates or functional coatings. The results on temperature-induced assembly should help improving electrically or thermally conductive pastes and aid the development of responsive materials, as the particle mobility and hence the adaptability of the bulk material can be controlled via temperature. The results also pose several fundamental questions that justify further investigations. In particular:

Chapter 2—Large-Area Convective Assembly. Large-area two-dimensional colloidal crystals of charge-stabilized particles can be produced by convective assembly. The shape of the meniscus in which the assembly takes place determines the robustness of the deposition against perturbations and hence the quality of the crystals. The withdrawal rate of the substrate plays a key role in shaping the meniscus. It was found that at withdrawal rates above $\approx 50\,\mu$/s a very thin sub-meniscus was extracted (Fig. 6.1a). According to the results of this section, such a flat meniscus should lead to most perfect films. However, extremely high particle concentrations are needed for film deposition and additionally the substrate withdrawal becomes similar to the velocity of the particles measured in Chap. 3. The assembly of the particles in such a flat-meniscus configuration might differ significantly from the presented mechanisms.

This work did not focus on the crystallography of the particle films. The relation between crystallographic axes and deposition direction, the formation of crystal defects, the motion of dislocations in the crystalline packings, the growth, decay and sudden complete reorientation of single-crystalline domains observed during deposition provide interesting future experiments.

The strongest limitation to the film quality are presently drying cracks. Minimizing the distance between the particles during the initial assembly by adjusting the particle stabilization will hinder additional rearrangement under the action of capillary interactions.

Finally, the presented models for convective particle assembly rely on a meniscus confining the particle during assembly. Using very fine particles with diameters smaller than the height of the flat adsorbate film preceding the main meniscus may lead to a qualitatively different deposition and assembly characteristics.

Chapter 3—Convective Crystallization. Two mechanisms compete in the crystallization in convective assembly, convective steering and immersion forces. The results indicate that these two mechanisms assure crystallization at all temperatures and particle sizes. However, fine nanoparticles were only used in preliminary experiments not presented in this work. Limited packing quality was found, with pronounced dependency of the crystallinity on the particle material. Again, the

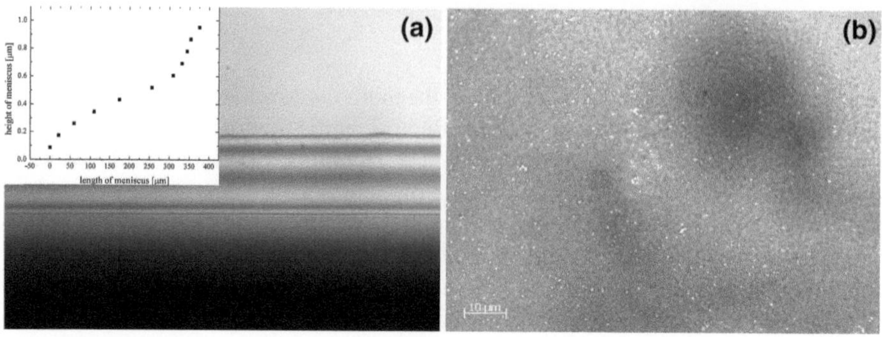

Fig. 6.1 Promising future directions in convective particle assembly: **a** Withdrawing the substrate with velocities around $100\,\mu$m/s generates an extremely flat sub-meniscus with surface slopes of $1\,$nm/μm (inset shows height measurement). Such a meniscus shape should be optimal for particle deposition, but requires very high particle concentrations. **b** Lowering the pH-value in the suspension reduces the ζ-potential of the particles. At pH 4 amorphous particle packings of 500 nm polystyrene spheres are deposited with convective assembly

assembly mechanisms in suspensions of small particles combined with the effect of the particle material and particle stabilization should be investigated further.

In analogy to the temperature-induced particle assembly, contact mechanics among the particles provide the tool to switch between amorphous and crystalline packings. The repulsive barrier must be removed to encounter contact among the particles. Precise tuning of the pH could provide an interaction potential in which the particles do not agglomerate in the suspension, but get into contact only in the assembly region. Preliminary results show the feasibility of such a route to deposit amorphous particle films (Fig. 6.1b).

Chapter 4—Temperature-induced Agglomeration. Alkyl thiol-stabilized particles undergo a glass transition upon cooling. The different growth kinetics in the reaction-limited and the diffusion-limited regimes allow production of morphologies that range from filamentous structures on a single particle level and globular structures to filamentous superagglomerates. Crystallization is effectively suppressed. It would be interesting to evaluate whether the observed mechanisms can be found using other stabilizing ligands, especially unsatured alkenyl ligands, which are commonly used for nanoparticle stabilization.

The genesis of the different morphologies requires further experiments. Up to now the kinetics are assumed to depend on the state of the ligand shell and to be independent of particle concentration. However, particle concentration will affect nucleation and growth of agglomerates; thus, size and shape of the agglomerates will depend on the particle concentration. Properties like mechanical stability or conductivity may vary for different morphologies.

Chapter 5—Temperature-induced Crystallization. Formation of crystalline packings of alkyl thiol-stabilized particles requires precipitation of the particles at elevated temperatures. The interactions of the ligand chains, best represented by the melting point of the ligands, determine the temperature needed for crystallization. Below the

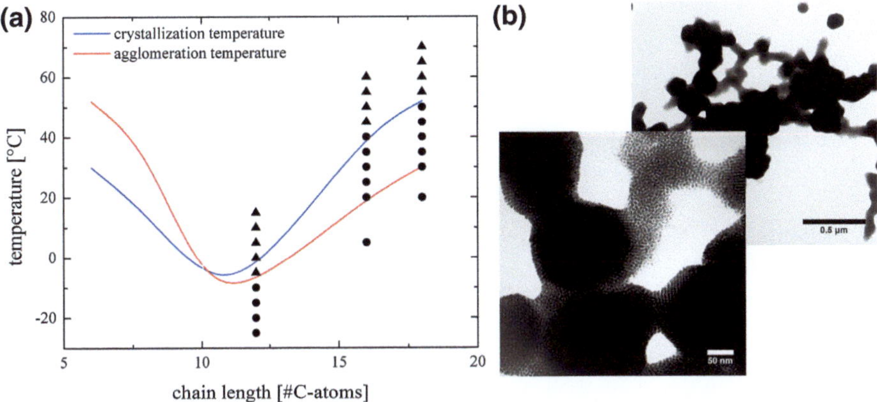

Fig. 6.2 Promising future directions in temperature-induced particle assembly: **a** A hypothetical chart of agglomeration and crystallization temperatures versus ligand chain lengths. Triangles denote experimental determined temperatures at which crystals formed upon crystallization, circles denote amorphous agglomeration. The *red line* indicates the evolution of the agglomeration temperature upon cooling. Above 12 C-atoms, the agglomeration temperature increases due to ligand-ligand interactions. Below 12 C-atoms it may increase again due to unshielded core-core-interactions. The *blue line* indicates the temperature required to induce crystalline agglomeration. Below 12 C-atoms it may increase slower than the agglomeration temperature, as the ligands stay molten. A crystallization temperature below the agglomeration temperature will allow for crystalline temperature-induced agglomeration. **b** Indications of spinodal decomposition leading to crystalline packings included in amorphous filaments in a C12 sample. Agglomeration was induced by 1-propanol at 20 °C

melting point of the ligands the particle have poor mobility within agglomerates. Particles of different materials and with different ligands should be considered for future experiments to test the universality of these results.

Particularly interesting would be the behavior of particles with shorter chains. Figure 6.2a summarizes the experimental findings for the ligand-dominated particles. At the C12-ligand length, the agglomeration line and the crystallization line nearly meet. As the ligand length gets shorter, the ligands should not shield the attractions of the cores anymore, and the suspension should behave like a VAN DER WAALS-gas with temperature independent interactions. The agglomeration temperature should increase with shorter ligands. At the same time, shorter ligand chains will stay liquid at lower temperatures. This implies that the agglomeration kinetics and the particle packing become fully separated. In such a regime spinodal-like agglomerates with crystalline packings might become feasible, as indicated by agglomerates from precipitation experiments with C12 particles (Fig. 6.2b).

Combining particles with different ligand length could be a versatile approach to the design of new structures. As an example, cooling a combination of particles with long ligand chains and particles with short ligand chains may initially lead to forming an amorphous spinodal backbone on which the particle with short chains could condense with further cooling, forming a crystalline coating.

Curriculum Vitae

Dr. Philip G. Born studied physics at the TU Dresden, Germany, and at the Uppsala University, Sweden. In 2007 he graduated from the Department for Materials Chemistry at Uppsala University. He joined the Chemical Vapor Deposition group at the Leibniz Institute for New Materials in Saarbrücken, Germany, where he worked on photocatalytic activity of core-shell nanowire structures. In 2008 he joined the group 'Structure Formation on Small Scales', where he worked on the self-organization and crystallization of colloidal particles, which is the content of this thesis. In 2012 he joined the Institute for Materials Physics in Space at the German Aerospace Center. There he focuses on the development of new light scattering methods for granular matter and the design of experiments on fluids, colloids and granular matter under microgravity.

P. G. Born, *Crystallization of Nanoscaled Colloids*, Springer Theses,   127
DOI: 10.1007/978-3-319-00230-9, © Springer International Publishing Switzerland 2013

Index

P. G. Born, *Crystallization of Nanoscaled Colloids*, Springer Theses,
DOI: 10.1007/978-3-319-00230-9, © Springer International Publishing Switzerland 2013